Spectroscopic Tools and Techniques for Analysis of Dental Materials

Spectroscopic Tools and Techniques for Analysis of Dental Materials: Current Trends introduces the dental materials and spectroscopic techniques applied for the analysis of such materials. The following individual chapters are primarily based on particular spectroscopic techniques and their applications. Techniques covered in this volume include X-ray spectroscopy, ultraviolet and visible spectroscopy, Fourier transform infrared spectroscopy, Raman spectroscopy, and mass spectrometry. This is an ideal book for dental professionals, researchers, and students interested in the analysis of dental materials.

Key Features:

- Individual chapters include brief introductions of specific techniques.
- Mathematical details are kept at a necessary minimum level.
- Includes case studies to suit the target audience.

Dr. Ashutosh Kumar Shukla has obtained his B.Sc., M.Sc., and D. Phil. degrees from University of Allahabad, India. His current research interest is primarily based on the applications of spectroscopy techniques in different fields of societal interest. In addition to his exposure to the continuous wave ESR spectrometers at leading institutes of his own country, he is enriched with hands-on experience with modern CW and pulsed ESR spectrometers at several leading departments such as the University of St. Andrews, Scotland, and Kazan State University, Kazan. Dr. Shukla is currently Professor of Physics at Ewing Christian College, Prayagraj, a constituent college of the University of Allahabad, India. He has been awarded the Bilateral Exchange fellowship and Visiting fellowship of Indian National Science Academy, New Delhi, in addition to research fellowships of Council of Science and Technology, Lucknow, and Council of Scientific and Industrial Research, New Delhi. With more than 22 years of experience of physics teaching, he has to his credit funded research projects, journal publications, textbooks, and more than 30 edited volumes published by reputed publishers.

Spectroscopic Tools and Techniques for Analysis of Dental Materials

Current Trends

Edited by
Ashutosh Kumar Shukla

CRC Press
Taylor & Francis Group
Boca Raton London New York

CRC Press is an imprint of the
Taylor & Francis Group, an **informa** business

Designed cover image: © Shutterstock

First edition published 2025
by CRC Press
2385 NW Executive Center Drive, Suite 320, Boca Raton FL 33431

and by CRC Press
4 Park Square, Milton Park, Abingdon, Oxon, OX14 4RN

CRC Press is an imprint of Taylor & Francis Group, LLC

ISBN: 978-1-032-82582-3 (hbk)
ISBN: 978-1-032-82885-5 (pbk)
ISBN: 978-1-003-50680-5 (ebk)

DOI: 10.1201/9781003506805

Typeset in Minion
by SPi Technologies India Pvt Ltd (Straive)

Contents

Contributors

Zühre Akarslan
Faculty of Dentistry Department of Dentomaxillofacial Radiology, Gazi University
Ankara, Türkiye

Associate Prof. Gülsün Akay
Faculty of Dentistry Department of Dentomaxillofacial Radiology, Gazi University
Ankara, Türkiye

Nuray Bağcı
Faculty of Dentistry Department of Dentomaxillofacial Radiology, Gazi University
Ankara, Türkiye

Agustina Ari Murti Budi Hastuti
Department of Pharmaceutical Chemistry, University of Gadjah Mada
Yogyakarta, Indonesia

Siti Nurul Hidayah
Department of Pharmaceutical Chemistry, University of Gadjah Mada
Yogyakarta, Indonesia

Gunawan Indrayanto
School of Medicine, Ciputra University
Surabaya, Indonesia

Tatang T. Irianti
Department of Pharmaceutical Chemistry, University of Gadjah Mada
Yogyakarta, Indonesia

Nebiha Gozde Ispir
Faculty of Dentistry Department of Dentomaxillofacial Radiology, Gazi University
Ankara, Türkiye

S. S. Sai Karthikeyan
Department of Periodontology, Gitam Dental College & Hospital
Visakhapatnam, Andhra Pradesh, India

Veena Koul
Centre for Biomedical Engineering, Indian Institute of Technology-Delhi (IIT-D)
New Delhi, India

Anisha Mishra
Department of Dentistry,
All India Institute of
Medical Sciences (AIIMS)
Bhubaneshwar, Odisha,
India

Eka Noviana
Department of Pharmaceutical
Chemistry, University of
Gadjah Mada
Yogyakarta, Indonesia

Umut Pamukcu
Department of Oral, Dental,
and Maxillofacial Radiology,
Gazi University
Ankara, Türkiye

Ilkay Peker
Faculty of Dentistry Department
of Dentomaxillofacial
Radiology, Gazi University
Ankara, Türkiye

Preeti Pandey
Department of Chemistry,
Kalinga University
Naya Raipur, Chhattisgarh,
India

Halida Rahmania
Department of Pharmaceutical
Chemistry, University of
Gadjah Mada
Yogyakarta, Indonesia

Anju Rana
Department of Periodontology,
Manav Rachna Dental College,
School of Dental Sciences,
Manav Rachna International
Institute of Research and Studies
Faridabad, Haryana, India

Abdul Rohman
Department of Pharmaceutical
Chemistry, University of
Gadjah Mada
Yogyakarta, Indonesia

Elif Sadik
Department of Oral and
Maxillofacial Radiology,
Ordu University
Ordu, Türkiye

Jitendra Sharan
Department of Dentistry,
All India Institute of
Medical Sciences (AIIMS)
Bhubaneshwar, Odisha, India

Manoj Pratap Singh
Advance Instrumentation
Research Facility, Jawahar Lal
Nehru University (JNU)
New Delhi, India

Meryem Toraman
Department of Oral and
Maxillofacial Radiology, Gazi
University
Ankara, Türkiye

Preface

I AM HAPPY TO PRESENT this collection before you. This collection covers the application of different spectroscopic techniques for the analysis of dental materials. This volume has 12 chapters. Chapter 1 highlights different spectroscopic techniques applied for dental material analysis and introduces the subject matter. Working principles of different spectroscopic techniques have been included in this chapter to acquaint you with the basics. Dental materials and their properties have been discussed in the next chapter. The following chapters talk about the different spectroscopic techniques and their applications for the analysis of such materials in detail. These chapters include the descriptions of spectrometers and experimental arrangements wherever appropriate to bring the sense of completeness. Infrared spectroscopy, Fourier transform infrared spectroscopy, ultraviolet and visible spectroscopy, X-ray spectroscopy, and Raman spectroscopy have been covered in these chapters. Mass spectrometry applications have been covered in two chapters. One of these two chapters illustrates how mass spectrometry can be integrated with chromatography. Applications of both gas chromatography-mass spectrometry (GC-MS) and liquid chromatography-mass spectrometry (LC-MS) have been covered in this chapter. There is a specific chapter that deals with the applications of spectroscopic methods to determine the toxicity of biomaterials used in dentistry. While going through a particular chapter, you will have a feel for the pedagogic approach. Mathematical details have been kept at the necessary minimum level. Case studies have been included to bring in an enhanced learning experience.

My heartfelt thanks to Mrs Rebecca Hodges-Davies, Commissioning Editor for Physics, CRC Press, Taylor & Francis Group, for giving me an opportunity to present this volume. I also thank Dr Danny Kielty,

Commissioning Editor for Physics, CRC Press, Taylor & Francis Group, and Sanika Shah, Editorial Assistant, Physics, CRC Press, Taylor & Francis Group, for their constant support during different stages of publication. My sincere thanks are also due to the contributors and reviewers who have enriched this volume through their valuable contributions and helped me to present quality content before you.

Ashutosh Kumar Shukla
Prayagraj

CHAPTER 1

Spectroscopic Analysis of Biomaterials Used in Dentistry

Anju Rana and S. S. Sai Karthikeyan

1.1 INTRODUCTION

The materials used in the oral cavity, where they directly interact with tissues, require specific physicochemical and biological properties. These properties include biocompatibility, ensuring safety locally and systemically, resilience against oral physicochemical factors, and biophysical inertness. Additionally, these materials need to accurately replicate natural tooth shapes, maintain stable mechanical properties, facilitate easy processing, provide satisfactory aesthetics, and remain cost-effective.

After placement in the oral cavity, the initial observed phenomenon is the formation of a biofilm on the material's surface. Understanding the material's structural and chemical components is crucial to prevent microleakage and biofilm development. Parameters such as surface structure, roughness, chemical composition, and reactivity must be evaluated when assessing materials for dental applications. These surface characteristics are crucial for understanding the material's mechanical and chemical behaviours in oral conditions.

Recent advancements in microscopic and spectroscopic techniques have enabled in-depth studies of the chemical structure and surface phenomena of these materials. Microscopy allows for high-magnification observation of surfaces, structural changes within materials, and intramaterial transformations. Techniques such as optical microscopy, electron

DOI: 10.1201/9781003506805-1

microscopy (EM), X-ray microscopy, and scanning probe microscopy facilitate advanced experimental research. Coupled with robust data acquisition and processing methods, these techniques enable comprehensive analysis of dental biomaterials, providing insights into their morphological, structural, and chemical properties. Microscopic methods also yield quantitative and qualitative information on chemical composition, employing techniques such as energy dispersive X-ray spectroscopy (EDS) for elemental mapping and electron backscatter diffraction (EBSD) for crystal orientation analysis.

In contrast, spectroscopy encompasses techniques based on the interaction of electromagnetic radiation with matter, often involving the generation and interpretation of spectra. Spectroscopic methods are advantageous in dental material studies due to their non-destructive nature and minimal sample requirements. Techniques such as infrared (IR) spectroscopy, Fourier transform infrared (FT-IR) spectroscopy, Raman spectroscopy, ultraviolet and visible (UV-Vis) spectroscopy, X-ray spectroscopy, and mass spectrometry (MS) play pivotal roles in analysing dental materials.

The diverse array of scientific methods employed today reflects the various hypotheses and challenges addressed by the research community. One such crucial challenge involves understanding the origin and complexity of interactions between biomaterials. Extensive studies across various biomaterials have shown that surface properties profoundly influence their interactions with the environment, dictating their short- and long-term behaviours. Effective methods for studying material interfaces ideally provide qualitative and quantitative information, are non-destructive, compatible with diverse substrates, and are straightforward in terms of usability and interpretation. While no single method meets all these criteria, spectroscopic techniques represent a category that fulfils many requirements. Over the decades, these methods have been extensively applied in fields, including analytical chemistry, biophysics, organic chemistry, and material science, supporting applications ranging from agriculture to clinical diagnostics.

Spectroscopy, defined by the International Union of Pure and Applied Chemistry (IUPAC) as the study of physical systems through their interaction with electromagnetic radiation, encompasses a wide range of techniques known for their versatility and reliability. Advances in spectroscopic techniques have significantly benefited dental biomaterials and dental implantology, a field bridging biomaterial science and biophysical chemistry. Significantly, there have been numerous studies conducted on dental biomaterials and dental implants. However, before utilizing any biomaterial or implant, its interaction with tissues must be evaluated to ensure long-term success.

1.2 BIOMATERIAL DEFINITION

The "Consensus Conference" held in Chester, UK, in 1991, is known for establishing a widely accepted definition of a biomaterial. It defined biomaterial as follows:

> [A]ny substance or combination of substances, other than drugs, synthetic or natural in origin, which can be used for any period of time, which augments or replaces partially or totally any tissue, organ or function of the body, in order to maintain or improve the quality of life of the individual.[1]

Here's a breakdown of the key components of the definition:

1. **Nature of the Substance**: It refers to any substance or combination of substances, excluding drugs, that can be used to augment or replace tissues, organs, or functions of the body.
2. **Purpose**: The primary purpose is to maintain or improve the quality of life of the individual using the material.
3. **Scope**: The definition encompasses both synthetic and natural substances used for medical purposes.

This definition outlines the characteristics that qualify a product as a medical, emphasizing its role in enhancing or substituting bodily functions to benefit the patient's quality of life.

Dr. Cohen, an orthopaedic surgeon, categorized biomaterials into four groups: metals, bone and bone derivatives used as grafts, plastics, and ceramics and composites.[2]

If classified on the basis of tissue-biomaterial response, biomaterial can be classified as bioinert, bioresorbable, and bioactive.[3]

1. **Bioinert**: Biomaterials classified as bioinert are characterized by their ability to remain relatively inert or non-reactive when implanted in the body. They do not elicit a significant immune response or induce a strong biological reaction from the surrounding tissues. Examples include certain metals like titanium and ceramics like alumina.

2. **Bioresorbable (or biodegradable)**: Biomaterials classified as bioresorbable are designed to degrade over time in the body. They are typically used for temporary applications where the material is gradually broken down and absorbed by the body's natural processes. This category includes polymers such as polylactic acid (PLA) and polyglycolic acid (PGA) used in sutures or implants that are meant to be replaced by natural tissue over time.

3. **Bioactive**: Bioactive biomaterials are designed to elicit a specific biological response at the interface with the body. They can promote cell adhesion, growth, and differentiation, leading to integration with surrounding tissues. Bioactive materials often have surface properties that facilitate interactions with biological molecules and cells. Examples include hydroxyapatite-coated implants used in orthopaedic applications to enhance bone growth.

These classifications help guide the selection and design of biomaterials based on their intended applications and desired interactions with the body's tissues.

1.3 PROPERTIES OF IDEAL BIOMATERIAL[4]

Various properties of ideal biomaterial are depicted in Figure 1.1.

1. Biocompatible (Non Carcinogenic and Non-Inflammatory)
2. Bioinert (No Adverse Response in Body)
3. Adequate Stability and Safety
4. Adequate Physical and Mechanical Properties
5. Bioactive
6. Cost-Effective and Easy Fabrication

1.4 APPLICATION OF BIOMATERIALS

Biomaterials play a crucial role in medicine and dental field, particularly in situations where tissues are damaged or diseased beyond natural repair. Some of the primary reasons for utilizing biomaterials are as follows[3,5]:

1. **Replacement of Damaged Tissues**: When tissues are damaged due to trauma, infection, cancer, or other pathological processes,

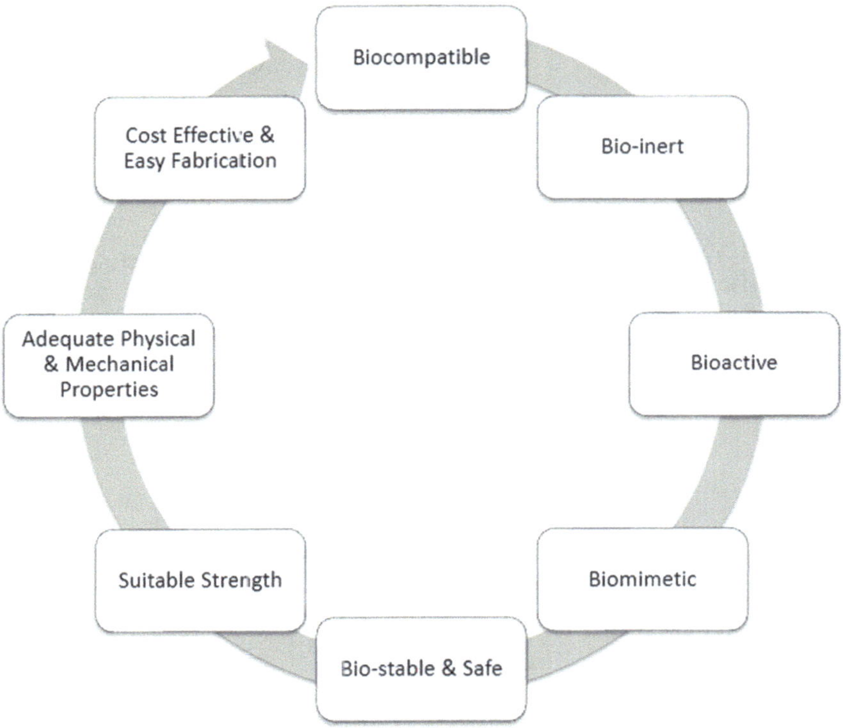

FIGURE 1.1 Characteristics of an ideal biomaterial.

biomaterials can be used to replace or repair the damaged tissue. For example, artificial joints made from biomaterials like titanium are commonly used to replace damaged hip or knee joints.

2. **Restoration of Function**: Biomaterials can restore lost function to tissues or organs that have lost their function due to disease or injury. This can improve a patient's quality of life significantly.
3. **Structural Support**: Biomaterials can provide structural support to tissues or organs that have been weakened. For instance, meshes made from biomaterials are used in hernia repairs to provide support to weakened abdominal walls.
4. **Biocompatibility**: Biomaterials are designed to be compatible with biological systems, minimizing adverse reactions when implanted in the body. This is crucial to ensure that the body accepts the biomaterial without causing inflammation or rejection.

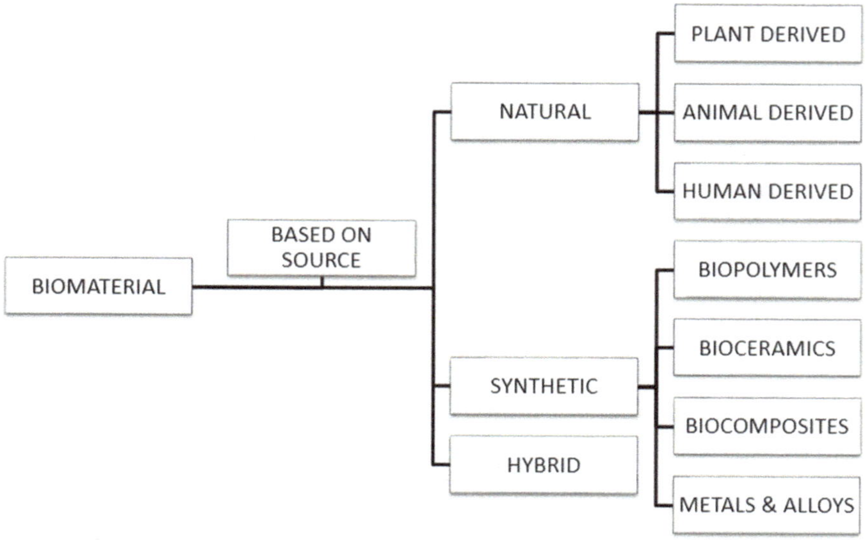

FIGURE 1.2 Classification of Biomaterials.

5. **Promotion of Healing**: Some biomaterials are designed to promote healing processes in the body. For example, scaffolds made from biomaterials can support the growth of new tissue in regenerative medicine approaches.

Overall, biomaterials from various sources such as natural, synthetic, or hybrid source (Figure 1.2) offer versatile solutions in medical applications by providing alternatives to damaged or diseased tissues, thereby improving patient outcomes and quality of life.

1.5 TYPES OF BIOMATERIALS AVAILABLE

1.5.1 Orthopaedic Prosthesis/Implants

Biomaterials play a crucial role in orthopaedics, particularly in the advancement of implant technology. Conditions such as osteoarthritis and rheumatoid arthritis can severely compromise the function of mobile joints such as the hip, knee, shoulder, ankle, and elbow. The pain associated with weight-bearing joints like the hip and knee can be debilitating, significantly limiting mobility. Since the advent of anaesthesia, antisepsis, and antibiotics, the replacement of these joints with prostheses has become feasible. This approach not only relieves pain but also restores mobility, providing immense benefits to a large number of patients globally.

1.5.2 Cardiovascular Applications

In the cardiovascular system, which includes the heart and blood vessels responsible for circulating blood throughout the body, issues can arise with heart valves and arteries. Heart valves can undergo structural changes that hinder their full opening or closing, and these damaged valves can be replaced with various substitutes. Similar to orthopaedic implants, materials such as ceramics, metals, and polymers are utilized in their construction. Arteries, especially the coronary arteries and those in the lower limbs, can become obstructed by fatty deposits (atherosclerosis). In some instances, segments of these arteries can be replaced with artificial counterparts.

1.5.3 Ophthalmics

Eye tissues are susceptible to various diseases that can diminish vision and potentially result in blindness. For instance, cataracts cause opacity of the lens, which can be remedied by replacing it with a synthetic polymer intraocular lens. Biomaterials are also employed in contact lenses, which intimately interact with eye tissues. Similar to intraocular lenses, contact lenses serve the purpose of maintaining and restoring vision.

1.5.4 Dental Applications

In the oral cavity, bacterial diseases can lead to the destruction of both teeth and the supportive gum tissues. Dental caries, which involve the demineralization and breakdown of teeth due to bacterial activity in plaque—a sticky film that harbours bacteria on tooth surfaces—can result in significant tooth loss. Various materials can be used to replace or repair entire teeth or sections of teeth affected by these conditions.

1.5.5 Wound Healing

One of the earliest applications of implantable biomaterials dates back to the introduction of sutures for closing wounds. As early as 2000 BC, ancient Egyptians utilized linen as a suture material. Modern synthetic sutures encompass a variety of materials, primarily polymers, which are the most widely used, along with certain metals such as stainless steels and tantalum.

Another critical category in wound healing involves fracture fixation devices, which encompass bone plates, screws, nails, rods, wires, and similar devices used in fracture treatment. While nonmetallic materials like carbon–carbon composites have been explored, the majority of fracture fixation devices employed in orthopaedic applications are made from metals, particularly stainless steels.

1.5.6 Drug-Delivery Systems

One of the most rapidly advancing fields in implant applications involves devices designed for controlled and targeted drug delivery. Numerous efforts have been made to integrate drug reservoirs into implantable devices to achieve sustained and ideally controlled release. Some of these innovations utilize novel polymeric materials as carriers for drug delivery.

1.6 TYPES OF DENTAL BIOMATERIALS AVAILABLE

1.6.1 Polymers

Today, polymers are extensively utilized in dentistry, benefiting from advancements in material science that have enabled the development of materials with properties akin to natural biomaterials. A prominent application is the use of polymeric base materials in both complete and partial dentures. Additionally, polymers are integral to denture soft liners, resin cements, and pit and fissure sealants. Some of the polymers used for denture bases include vulcanite, celluloid, phenol formaldehyde (Bakelite), polyvinyl chloride (PVC), and poly methyl methacrylate (PMMA).

Despite the introduction of new materials for denture bases, PMMA remains the preferred choice for both types of dentures. The advantages that have established PMMA as the most commonly used polymer include its affordability, lightweight nature, aesthetic appeal, low solubility, minimal water absorption, ease of repair, and straightforward processing. However, a weakness of this polymer is its susceptibility to clinical failure and fracture from impact forces or fatigue. Efforts have been made to address this issue by enhancing the flexibility and strength of PMMA through the incorporation of substances such as copolymers, various glass and metallic fibres, and nanoparticles.

Another application of polymers is seen in tissue engineering and the regeneration of dental structures. Dental implants, which are widely used to replace missing teeth, require a sufficient amount of bone in the tooth socket to prevent biological and mechanical complications. In a study involving eight human cases, researchers discovered that the use of poly(lactic-co-glycolic acid) (PLGA) prior to placing metallic implants can enhance and repair the alveolar socket.

Polymers also serve a crucial role in tissue engineering as delivery vehicles. By integrating regenerative factors, such as growth factors, with polymers, these substances can be transported to specific locations and released gradually to support sustained regeneration.

1.6.2 Ceramics

Ceramics are composed of metallic elements combined with non-metallic materials like oxides, nitrides, and silicates. They exist in two forms: crystalline solids and amorphous solids, with the latter group also known as glass.

Ceramics exhibit hardness, stiffness, and brittleness due to their ionic and covalent inter-atomic bonding. They possess excellent resistance to compressive stresses but are susceptible to shear and tensile stress. Their translucency or opacity depends on their microstructure, with higher amounts of amorphous phases leading to greater translucency and higher amounts of crystalline phases resulting in increased opacity.

In dentistry, ceramics find various applications. They are used as fillers in composite resins, glass ionomers, investment materials, and porcelain cements.

Porcelains are dental restorations crafted in the laboratory for applications like crowns, bridges, and veneers. These materials offer several advantages, such as superb biocompatibility, natural chemical composition, aesthetic appeal, resistance to dissolution, and high durability.

1.6.3 Metals

Currently, metal-free dental technologies have advanced rapidly; however, metals are still debated for their durability and long-term clinical applications.

Metals are elements that ionize to form positively charged ions in solution, comprising approximately 70% of the periodic table. Upon ionization, metals release their valence electrons. In dentistry, the behaviour and utilization of metals depend on their ability to generate stable and positively charged ions.

Metallic elements used in dentistry are categorized into two groups: noble and base metals. Base metal elements oxidize when heated in open air and are typically more economical compared to precious metals. This group includes titanium, silver, copper, zinc, indium, tin, gallium, nickel, and cobalt.

Gold, platinum, palladium, iridium, rhodium, osmium, and ruthenium are classified as noble metals due to their exceptional resistance to oxidation, tarnishing, and corrosion during heating, soldering, and oral use.

Gold alloys are among the metal alloys commonly employed in modern dentistry, predominantly composed of noble metals such as gold, platinum, and palladium. Gold has a long-standing history of use in dentistry and continues to be utilized across various dental applications despite the introduction of new materials. Its key attributes include high biocompatibility, minimal corrosion, reduced plaque accumulation, and excellent marginal fit.

Their primary applications in dentistry encompass are as follows:

1. Indirect restorations (inlay and onlay)
2. Post and core
3. Crown (all metal and porcelain fused to metal [PFM])
4. Telescopic crown that is used as retentive element in over-denture.
5. Bridge
6. Framework for removable denture
7. Implants

The utilization of gold alloys in PFM restorations has evolved significantly over the past 20–30 years. This is attributed to the metal's distinctive properties, including its resistance to bending and a thermal expansion coefficient closely matching that of porcelain. Apart from dental restorations, certain orthodontic wires and splints can also be crafted from gold alloys.

1.6.4 Composites

Composite resins are employed in dentistry for reconstructing tooth structure, altering colour, correcting contours, and enhancing the aesthetic appearance of tooth surfaces. These composites typically consist of three primary components: an organic resin matrix, inorganic filler components, and a coupling agent, alongside other additives.

The most commonly used resins include dimethacrylates (such as Bis-GMA) and urethane dimethacrylate oligomers (UDMAs). These resins are viscous fluids to which low molecular weight monomers like triethylene glycol dimethacrylate (TEGDMA) are added to adjust their consistency for handling. Modern materials feature carbon–carbon double bonds that undergo polymerization upon reaction, converting the material into a polymer. New monomers like silorane are designed to minimize shrinkage and internal stresses caused by polymerization, thereby enhancing the durability of dental restorations.

Composites are categorized based on the size and type of filler components they contain. Size-wise, composites are divided into two groups: fine fillers (with an average size of 0.4–3 micrometres) and micro-fine fillers (with an average size of 0.2–0.4 micrometres). In terms of filler type, composites are classified into two groups: microhybrid and microfilled. Microhybrid composites incorporate a combination of fine and micro-fine filler particles, with micro-fine particles enhancing packing and overall properties. Microfilled composites consist predominantly of micro-fine filler particles that contribute to a large surface area. Coupling agents may also be utilized to ensure a strong bond between the inorganic fillers and the organic resin matrix.

Composites offer several advantages such as aesthetic appeal, bonding to tooth structure through micromechanical adhesion, ease of repair, minimal need for extensive tooth surface preparation compared to amalgam, reduced environmental mercury release, decreased mercury exposure for dentists, and lower risk of corrosion.

Some drawbacks of composites include polymerization shrinkage, potential for secondary caries due to marginal leakage, requirement for advanced skills and high technique sensitivity, time-intensive application, necessity for meticulous isolation and dry conditions during restoration placement, risk of chipping, and lower toughness compared to alternative materials.

1.7 VARIOUS TECHNIQUES TO VISUALIZE BIOMATERIAL–TISSUE INTERACTION[6]

1.7.1 The Need to Visualize Biomaterial–Tissue Interactions

Biomaterials are increasingly integral in biomedicine, facilitating tasks like scaffolding for engineering functional tissues in regenerative medicine and transporting bioactive agents in precision medicine. Achieving optimal biomaterial performance necessitates non-invasive, precise visualization of biomaterial–tissue interactions. Historically, imaging these interactions has been challenging due to the need for engineered imaging contrast in biomaterials or cells, compounded by complexities like increased imaging depth and interference from biological systems in vivo. Addressing these challenges has involved coordinated efforts in imaging and materials sciences over recent decades, leading to significant advancements in biomaterials research. Modern biomedical imaging has been tailored to enhance spatial resolution, penetration depth, temporal response, detection sensitivity, and chemical specificity for imaging biomaterials. Concurrently, engineering strategies have evolved to impart novel contrast mechanisms to biomaterials and cells, facilitating easier characterization of biomaterial–tissue interactions.

Biomedical imaging modalities rely on interactions between probing energy formats (such as light, sound, magnetic fields, or X-ray photons) and biomaterials/tissues, involving absorption, scattering, and polarization of the probing energy. Various criteria categorize imaging technologies, including resolution, imaging depth, and contrast mechanism. Here, we focus on contrast mechanisms, crucial for selecting appropriate technologies based on the physical, chemical, and biological properties of materials or tissues. Common biomedical imaging modalities are classified by their contrast mechanisms into acoustic, magnetic, optical, electron, X-ray, and nuclear imaging. Each modality is discussed in terms of potential use for characterizing biomaterial–tissue interactions, organized by probing energy levels from low to high. Importantly, the probing energy level indicates potential damage, with higher energy necessitating greater caution, especially in studies involving sensitive cell types and dedicated biomaterials.

1.7.2 Acoustic Imaging of Mechanics

Ultrasound (US) imaging is widely utilized for its ability to offer three-dimensional (3D) anatomical details of engineered tissue constructs and newly developed tissues at depths reaching tens of centimetres. Due to its low acoustic wave energy, US imaging is recognized as one of the safest imaging techniques, characterized by high imaging speeds and deep penetration capabilities. Various US imaging methods have been employed to investigate the mechanical characteristics of biomaterials and their interactions with neighbouring tissues.

Ultrasound (US) imaging is particularly valuable for studying the mechanical properties of biomaterials, including parameters such as mass density, modulus, and cavitation. For instance, conventional B-mode US imaging has been utilized to evaluate mineralization changes in collagen hydrogels. Doppler US imaging tracks flow dynamics in engineered vascular grafts and vascularization of hydrogels in vivo. US elastography monitors scaffold degradation (changes in modulus) in vivo.

However, US imaging has two main limitations:

1. It lacks sensitivity to the functional and molecular characteristics of biomaterials and tissues (e.g., chemical composition and functional status) due to its mechanical contrast mechanism.
2. Its spatial resolution is limited and typically cannot reach single-cell levels because of the long acoustic wavelengths used to generate the images.

1.7.3 Magnetic Imaging of Water Content

Magnetic resonance imaging (MRI) detects water content or proton density in an object and has long been used to monitor the development of engineered tissues like adipose tissue, cartilage, and vasculature, using either endogenous or exogenous contrasts. MRI is also employed to track stem cell differentiation and degradation of engineered scaffolds and hydrogels, utilizing specific macromolecules labelled with functionalized superparamagnetic iron oxide nanoparticles. It is a non-invasive, non-ionizing imaging method widely applied in in vivo studies of tissue-biomaterial interactions in small animals, particularly at high magnetic fields (>7T) that offer improved spatial resolutions.

However, MRI is costly and not widely accessible to the general research community. Additionally, the potential effects of strong magnetic fields on normal cell functions and small animal behaviour require further conclusive demonstration.

1.7.4 Optical Imaging of Chemical Composition

Optical imaging has become indispensable in biomedical laboratories due to its adaptable imaging contrasts, high spatial resolutions, and user-friendly nature. It is particularly sensitive to the chemical information of objects. Techniques like confocal microscopy (CM) and multi-photon microscopy (MPM) are extensively utilized for imaging biomaterials and tissues both in vitro and in vivo, often employing fluorescence markers. Cell viability within 3D scaffolds can be evaluated using the intracellular autofluorescence of nicotinamide adenine dinucleotide (NADH).

Photoacoustic tomography (PAT) is effective for assessing tissue angiogenesis and oxygenation saturation of haemoglobin in vivo within hydrogel scaffolds. Optical coherence tomography (OCT), however, relies on scattering contrast and is not sensitive to chemical composition. OCT is valuable for evaluating 3D structural changes in engineered scaffolds due to biomaterial degradation and tissue matrix deposition.

Overall, optical imaging dominates the characterization of biomaterial–tissue interactions, but these techniques are limited by relatively shallow penetration depth due to strong optical attenuation by biological tissues and biomaterials.

1.7.5 X-Ray, Electron, and Nuclear Imaging of Material Structures at Scales

Electron, X-ray, and nuclear imaging utilize high-energy electromagnetic waves to characterize material structures across vastly different scales. These modalities are grouped together due to their high probing energy and potential for damaging imaged materials.

Conventional X-ray imaging, relying on X-ray attenuation, excels in detecting hard structures like engineered bones but has limited sensitivity to soft tissues and biomaterials in vivo. Phase-contrast X-ray imaging enhances sensitivity, particularly for imaging polymer scaffolds cultured or implanted in tissues, though it involves ionizing radiation, which can potentially harm samples.

EM, encompassing techniques such as scanning electron microscopy (SEM) and transmission electron microscopy (TEM), is pivotal for nanoscale characterization of biomaterials and cell-biomaterial interactions. EM offers unparalleled resolution for imaging nanoparticle dimensions, scaffold pore sizes, and fibre orientations. However, EM requires costly maintenance, offers shallow penetration depth, necessitates complex sample preparation, and has limited capability to image live cells.

Nuclear imaging, including positron-emission tomography (PET) and single-photon emission computed tomography (SPECT), tracks engineered tissues and cells labelled with radioactive tracers. Nuclear imaging benefits from deep tissue penetration and detection sensitivity at picomolar levels. PET commonly monitors cell metabolism using 18F-fluorodeoxyglucose (FDG), a radiative glucose analog. However, nuclear imaging's spatial resolution is compromised by tissue scattering of gamma rays and random annihilation events. Therefore, it is often combined with MRI or X-ray computed tomography (CT) for comprehensive structural and molecular information.

1.7.6 Engineering Properties of Materials and Tissues to Enable Visual Characterization

As previously discussed, although different imaging modalities can directly visualize specific properties of specimens, their sensitivity may not always suffice to adequately characterize biomaterial–tissue interactions. Therefore, it is frequently essential to introduce exogenous contrast mechanisms to the non-biological materials, the biological entities, or both, in order to enhance their visibility under the chosen imaging techniques.

1.7.7 Physicochemical Contrast Mechanisms

Typically, a single imaging modality can only capture specific properties of a sample at a time, making it challenging to simultaneously image both biomaterials and biological entities. However, if there is a significant difference in properties between the material and the cells, such as their response to contrast agents, they can still be distinguished. This differentiation is usually facilitated by introducing exogenous contrast mechanisms.

The choice of contrast mechanism depends heavily on the imaging modality being used. For mechanical imaging, such as ultrasound (US) elasticity imaging, contrasts are based on acoustic scattering and physical resonance properties. Biomaterials with high acoustic scattering properties, like dense polymer scaffolds, enable imaging of scaffold degradation or regeneration in engineered tissues. Alternatively, contrast agents like gas-filled microbubbles or biogenic nanostructures can enhance contrast by resonating under US waves.

In MRI, which relies on atomic relaxation, water-rich hydrogels similar to biological tissues can be imaged without additional contrast agents. However, distinguishing between scaffolds and cells within them often requires using magnetic nanoparticles as external labels for the cells.

Optical imaging modalities rely on various contrast mechanisms such as scattering, absorption, and fluorescence. Polymeric biomaterials and cell membranes scatter light, facilitating OCT imaging. Enhanced contrasts between cells and scaffolds can be achieved through labels with higher refractive indices like magnetic beads or gold nanocages. Fluorescent molecules are commonly used in CM and MPM, while PAT utilizes absorption contrasts, often enhanced by contrast agents like chromogenic dyes or gold nanoparticles.

CT imaging primarily relies on X-ray attenuation for imaging hard tissues such as bones, which possess intrinsic contrasts. Adapting existing CT contrast agents based on nanoparticles could potentially broaden CT imaging applications beyond bones to other tissue types.

EM provides ultra-high-resolution imaging but is limited to thin samples. SEM can image surfaces without external contrasts, while TEM requires heavy metal-containing dyes to enhance contrasts in ultrathin sections of polymers and cells.

Each imaging modality employs specific contrast mechanisms tailored to the properties of biomaterials and biological entities, often requiring exogenous contrast agents to optimize imaging outcomes.

PET imaging, in contrast to the previously mentioned modalities, has seen limited application in studying scaffold–tissue interactions, primarily due to its restricted accessibility and lower spatial resolution. An early instance involved the use of 18F-FDG, a typical positron-emitting contrast agent used in clinical settings, to label cardiomyocytes within a fibrin matrix for PET imaging. The potential use of 18F-FDG and other radioactive substances could theoretically extend to various other cell types and even the labelling of biomaterial scaffolds.

1.7.8 Engineered Biological Contrast Mechanisms

Efforts have been made to enhance contrast mechanisms for imaging by genetically engineering target cells, minimizing reliance on externally introduced contrast agents which may have low uptake efficiency and could potentially impact cell behaviour negatively. Optical imaging methods, such as fluorescence, luminescence, and chromogenic techniques, have been particularly advanced through these approaches. The green fluorescent protein (GFP), first isolated in the 1960s and widely used since the 1990s, is a prime example of genetically encoded fluorescent labels that have revolutionized biological imaging across various fields from molecular biology to tissue engineering. Bioluminescence, where light emission occurs without external excitation, has also found extensive use in visualizing engineered cells, driven by enzymatic reactions such as luciferase oxidation of luciferin.

Chromogenic methods, primarily relying on light absorption for contrast, have been traditionally limited to thin sample layers due to optical imaging's challenge in penetrating deeply and capturing volumetric data. However, PAT has recently overcome these limitations by enabling three-dimensional imaging of chromogenic contrast mechanisms engineered within cells. Examples include genetically encoded reporters like LacZ, which converts X-gal molecules into blue precipitates within cells, as well as photothermal conversion-enhanced fluorescent proteins with low extinction coefficients, tyrosinase-melanin, and chromoproteins facilitating direct chromogen production.

While these engineered biological contrasts are predominantly used in optical imaging due to their compatibility with optical properties, exploring methods like in situ synthesis of nanomaterials within biological cells through green chemistry—mimicking cellular metabolic processes—could open new avenues for other imaging modalities such as ultrasound (US), OCT, MRI, and EM. Extensive research is required to ensure the biocompatibility and effectiveness of these methods in biomedical imaging applications.

1.8 SPECTROSCOPIC TECHNIQUES FOR SURFACE ANALYSIS OF DENTAL BIOMATERIALS[7,8]

The rapid advancements in dental biomaterials are facilitated by continuous innovations in materials and research methodologies. With a wide array of ceramic, metallic, synthetic, and composite materials available for dental applications, it is crucial for dental professionals to carefully select materials based on their specific limitations and characteristics.

Dental biomaterials used in direct contact with oral tissues must meet stringent requirements regarding their physicochemical and biological properties. These include biocompatibility, safety for local and systemic health, resistance to oral physicochemical conditions, and biophysical inertness. Moreover, these materials must accurately replicate natural tooth shapes, maintain stable mechanical properties, facilitate easy processing, offer satisfactory aesthetics, and be cost-effective.

Upon introduction into the oral cavity, dental biomaterials immediately encounter the formation of biofilms on their surfaces. Understanding the structural and chemical composition of biomaterials is essential to prevent micro-leakage and biofilm formation. Surface characteristics such as structure, roughness, chemical composition, and reactivity play crucial roles in determining the material's performance under oral conditions and are closely related to its mechanical and chemical behaviours.

Recent advancements in microscopic and spectroscopic techniques enable detailed studies of the chemical structure and surface phenomena of dental materials. Microscopic techniques, including optical microscopy, EM, X-ray microscopy, and scanning probe microscopy, provide high-resolution observations of surfaces and intra-material changes. These techniques, combined with advanced experimental protocols and data processing, allow comprehensive analysis of dental biomaterial systems. They provide insights into morphological, structural, chemical composition, elemental mapping (energy dispersive X-ray spectroscopy—EDS), crystal orientation (EBSD), and other material characteristics.

In addition to microscopy, spectroscopic techniques are invaluable for studying dental materials. The focus of this chapter will primarily be on spectroscopic techniques utilized for analysing biomaterials in dentistry. Spectroscopy involves the interaction of electromagnetic radiation with matter to generate and interpret spectra, providing detailed information on chemical structures. Techniques such as IR spectroscopy, FT-IR spectroscopy, Raman spectroscopy, UV-Vis spectroscopy, X-ray spectroscopy, and MS are non-destructive and require minimal sample amounts, making them ideal for dental material analysis (Figure 1.3).

The integration of spectroscopic techniques with microscopic approaches enhances the understanding of dental biomaterial's chemical properties, enabling researchers to evaluate their performance and suitability for clinical applications effectively. These methodologies contribute significantly to advancing dental biomaterial science, ensuring the

SPECTROSCOPIC TECHNIQUE	ELECTROMAGNETIC SPECTRUM	SPECTRUM RANGE (WAVELENGTH)
Microwave spectroscopy	Microwave	$1\text{-}10^{-3}$
Infrared spectroscopy Raman spectroscopy UV Visible spectroscopy Atomic absorption spectroscopy	Infrared	$10^{-3} - 10^{-6}$
Fluorescence spectroscopy Phosphorescence spectroscopy X-ray diffraction X-ray fluorescence	Ultraviolet and Visible	$10^{-6} - 10^{-8}$
X-ray photoelectron spectrometry Mass spectrometry	X-ray	$10^{-9} - 10^{-12}$
Mossbauer spectroscopy	γ-ray	$10^{-12} - 10^{-14}$

FIGURE 1.3 Spectral range for different spectroscopic techniques (Adapted from *Materials (Basel)*. 2021;14(10):2624).

development of safer, more durable, and aesthetically pleasing dental restorations.

Typically, spectroscopic techniques are classified based on several criteria, including the range of electromagnetic radiation utilized, the properties of the systems being studied, and the method used to collect spectral data, which pertains to how energy is exchanged between the radiation and the matter. This classification according to the range of radiation is directly linked to the specific experimental methodologies employed. Different radiation sources, detectors, and dispersion devices are employed depending on the spectral range being investigated. For instance, in optical spectroscopy, dispersion devices such as prisms and diffraction gratings are commonly used to separate and analyse the spectral components of light.

Spectroscopy can be categorized based on intrinsic characteristics of the studied processes, distinguishing between nuclear, atomic, and molecular spectroscopy, with a specific focus on condensed systems. This classification is linked to the specific energy levels involved in the transitions of the studied systems.

Depending on the method of radiation measurement, three types of spectra are identified: absorption spectroscopy, emission spectroscopy (ES), and Raman spectroscopy. Absorption spectra encompass transitions from lower to higher energy levels, corresponding to energy absorption (photon absorption). The simplest absorption spectrum occurs when only the lowest energy level (ground state) is occupied, influenced by the system's temperature in thermodynamic equilibrium, typically assumed to be room temperature where only the ground state is filled.

Emission spectra involve transitions from higher to lower energy levels, resulting in energy reduction and photon emission. Raman spectra are characterized by changes in the frequency of scattered radiation (νr) relative to the frequency of incident radiation (νp).

1.9 INFRARED SPECTROSCOPY AND FOURIER TRANSFORM INFRARED SPECTROSCOPY[9–14]

1.9.1 Basic Working Principle

The infrared (IR) region of the electromagnetic spectrum spans from approximately 14,000 to 200 cm^{-1} in wavenumber (the reciprocal of wavelength), situated between the visible and microwave regions. In this range, the energy absorbed is sufficient to induce oscillations of chemical bonds within molecules but not strong enough to break these bonds. Molecules rotate around their symmetry axes while their atoms oscillate between equilibrium positions. IR absorption spectra are acquired by measuring the relative intensity of transmitted or absorbed radiation as a function of wavenumber, typically plotted as transmittance or absorbance versus wavenumber (cm^{-1}).

Various types of IR spectrophotometers exist, such as filter photometers, double-beam spectrophotometers, Fourier transform (FT) spectrometers, and attenuated total reflectance (ATR) FT-IR instruments. Modern FT devices illuminate the sample simultaneously with radiation spanning the entire IR range. After passing through the sample, the radiation undergoes interference, and the spectrum is derived by Fourier transforming the recorded interferogram using specialized software. This method provides information about vibrational modes of atomic groups within the dental material being analysed.

FT-IR spectroscopy is primarily used for qualitative analysis due to the characteristic nature of the FT-IR spectrum for each substance. Beyond identifying pure substances, FT-IR can detect additional substances in mixtures and analyse interactions between atomic groups of different compounds. Changes in the spectrum, such as additional peaks or shifts relative to the spectrum of the pure compound/material, indicate the presence of different functional groups or interactions within the mixture.

Other IR spectroscopic techniques include transmission spectroscopy (TS), internal reflection spectroscopy (IRS or ATR), external reflection spectroscopy (ERS), diffuse reflection spectroscopy (DRS), ES, and photoacoustic spectroscopy (PAS). Each of these methods offers unique advantages depending on the specific application and sample characteristics.

1.9.2 Type of Samples Tested

IR spectroscopy allows for the examination of substances in gaseous, liquid, and solid states. For gas samples, spectra are obtained by introducing the gaseous sample into a purged cuvette. Liquids can be analysed either in their pure form or as solutions. Liquid samples are typically placed between sodium chloride plates, requiring approximately 1–10 milligrams of the liquid substance. If a liquid dissolves sodium chloride, silver chloride plates can be used instead.

Solid samples are commonly tested in the form of suspensions and pellets or deposited as a glassy film on a substrate. To obtain the spectrum of a solid sample, it is often dispersed within an alkali halide pellet.

Alternatively, solid samples can be analysed by preparing a film from the solid material after evaporating the solvent from a droplet deposited onto a sodium chloride plate.

1.9.3 Characteristics

FT-IR spectroscopy enables the identification of characteristic vibrations associated with functional groups present in a compound of the material. Because the FT-IR spectrum is unique to each substance, this technique is primarily employed for qualitative analysis. Beyond distinguishing pure substances based on their FT-IR spectra, it allows for the investigation of additional substances and their interactions with specific functional groups of the original compound. Changes in the spectrum, such as the appearance of new peaks, indicate the presence of different functional groups or additional compounds. Shifts in peak positions relative to the original spectrum indicate interactions between specific groups of the original and added substances.

For instance, FT-IR spectroscopy was utilized to elucidate the structure of fibroin, a dental material, identifying the NH functional group characteristic of amino acids with a peak at 3309.80 cm^{-1}.[15] FT-IR measurements were also employed to obtain spectra of various bioceramic powders,[16] facilitating their identification based on absorbance patterns.

1.9.4 Advantages of Technique

IR is widely utilized for identifying functional groups and chemical compounds, both organic and inorganic, as well as assessing dental biomaterial compound purity. It offers cost-effectiveness, ease of operation, and rapid acquisition of IR spectra. Compared to other techniques, it is less time-consuming in terms of sample preparation, measurement, and result interpretation. Moreover, it is non-invasive and relatively straightforward.

In FT-IR spectroscopy, all frequencies are measured simultaneously, allowing spectra to be obtained within seconds, which is advantageous over conventional IR that requires longer measurement times. Another strength of IR is its capability to analyse small sample quantities, particularly enhanced by the ATR technique, which allows penetration into the sample to a depth of about 0.5–3 μm.

1.9.5 Limitations of Technique

However, a limitation of traditional IR spectroscopy lies in the fact that spectra are obtained by examining the absorption of monochromatic radiation (one wavelength) and scanning the sample by changing the wavelength gradually using a dispersion element like a prism or diffraction grating. Sample preparation is crucial, typically involving grinding the material, mixing it with potassium bromide (KBr) matrix, and compressing it into a tablet. Moisture in samples must be avoided as water strongly absorbs radiation in specific regions (around 3700 cm^{-1} to 1630 cm^{-1}), potentially masking the absorption bands of the substance being tested.

In the ATR sampling method, errors can arise from imperfect contact between the sample and the diamond crystal. Another drawback of FT-IR instruments is their single-beam configuration, whereas dispersive instruments usually feature a double-beam setup.

1.9.6 Applications

FT-IR spectroscopy was utilized to monitor the polymerization kinetics of dental resins and adhesives, aiming to enhance their mechanical properties.[17–21] Additionally, this technique was employed to investigate the impact of intermolecular collagen cross-linking on dentin's mechanical behaviour.[22] It was also used to assess the structural properties of heterocyclic compounds for potential use in pulp regeneration and to analyse biomimetic materials that replicate the mineral-organic complex of dentin and enamel.[23]

Furthermore, FT-IR spectroscopy served to characterize the structure of various dental materials, including implant materials, biopolymers, ceramics, resin nanocomposites, implant coatings, bioceramics, resins, cements, bioglass, and self-curing materials.[24–27]

1.10 RAMAN SPECTROSCOPY[28–31]

1.10.1 Basic Working Principle

Raman spectroscopy is a method employed to examine the vibrational and rotational spectra of molecules within a biomaterial. When light interacts with the sample, it scatters into components with frequencies different from the incident light, in addition to those with the same frequency (Rayleigh scattering). The resulting Raman spectrum depicts the intensity of scattered light plotted against frequency (wavenumber), known as the Raman shift, which is the difference between incident and scattered radiation frequencies.

There are two main techniques for obtaining Raman spectra: dispersive and FT methods. They differ in the wavelength of the incident light and detection methodology. The choice of laser wavelength is crucial, influencing the depth from which information is gathered in the sample surface.

Shorter wavelengths penetrate less deeply but provide surface details, whereas longer wavelengths may miss surface-related data, especially critical for multi-layer substrates or sensitive samples.

Advancements in Raman spectroscopy, such as resonance Raman spectroscopy (RRS), surface enhanced Raman spectroscopy (SERS), confocal Raman microscopy, and coherent anti-Stokes Raman spectroscopy (CARS), enhance sensitivity. Raman spectroscopy yields valuable insights into molecular geometric structure and chemical bonding nature. For crystalline samples, it provides data on structure, lattice arrangement, mechanical properties, stress, and phase transitions. Each dental biomaterial compound has a unique Raman spectrum akin to a fingerprint, enabling qualitative analysis, while quantitative data can be derived from signal intensities correlating with substance concentrations in the sample.

1.10.2 Type of Samples Tested

Samples can be analysed in diverse states such as solids, liquids, or vapours, at various temperatures, in bulk form, as microscopic particles or as surface layers. Common accessories used in Raman spectroscopy include holders for powder samples, cuvettes, holders for small liquid samples (such as nuclear magnetic resonance sample tubes), and clamps for irregularly shaped objects.

1.10.3 Sample Characteristics

In Raman spectroscopy, the intensity of radiation emitted by the sample correlates linearly with both its molar concentration and the thickness of the layer. This spectroscopic technique captures an oscillating-rotational spectrum that provides insights into the structure and chemical composition of the biomaterial under examination. Raman spectroscopy has become pivotal in characterizing 2D materials and phonon modes in crystals. It enables the extraction of properties such as the number of monolayers, inter-layer breathing and shear modes, in-plane anisotropy, doping effects, disorder, thermal conductivity, strain, and various phonon modes.

For instance, Raman mappings have confirmed that graphene nanocoatings on biomedical-Grade Ti-6Al-4V alloys exhibit remarkable structural stability, high resistance to mechanical stress, and chemical degradation, maintaining over 99% coverage even after undergoing corrosion processes.[32] In the field of dental materials, Raman spectroscopy is frequently employed to assess the degree of conversion (DC) in dental composites, adhesives, and the setting reactions of cements.[33]

1.10.4 Advantages of Technique

Similar to other spectroscopic methods, Raman spectroscopy offers short measurement times and straightforward sample preparation. Moreover, it is non-invasive, and the results obtained can be analysed quickly and easily.

1.10.5 Limitations of Technique

A significant drawback when using visible incident light in all spectrometer systems is the issue of fluorescence. This problem is more pronounced in the visible spectrum compared to the UV or near-IR regions. Raman scattering is a relatively weak phenomenon, necessitating a strong excitation source to achieve high power density in the biomaterial sample. Consequently, any trace impurities that exhibit fluorescence, in addition to fluorescence from the sample itself, can generate substantial interfering signals. Fluorescence occurs at energies lower than the excitation energy, often overlapping with the energy range covered by Raman Stokes scattering and potentially masking or distorting the Raman signal.

1.10.6 Applications of Technique

Raman spectroscopy has been applied to assess the polymerization rate of bioglass ceramics, detect and identify various compounds in implant materials, ceramics, and bioglass, analyse the phase composition of bioglass and ceramics, and evaluate the DC in resins and ceramics.[31,34–38]

1.11 ULTRAVIOLET AND VISIBLE SPECTROSCOPY[39,40]

1.11.1 Basic Working Principle

UV-V is spectroscopy is a long-standing analytical technique that relies on measuring electron spectra resulting from transitions between energy levels of valence electrons within the wavelength range of 200–750 nm. Absorption of energy in the UV-Vis range induces electron transitions from lower energy levels (π and n orbitals) to higher energy levels (anti-bonding π^* orbitals). This method offers insights into the molecular structure, presence of conjugated double bond systems, and aromatic systems of the compound under examination. The instrument employed for UV-Vis spectroscopy is the UV-Vis spectrophotometer.

1.11.2 Type of Samples Tested

Spectroscopic analysis is commonly conducted in solutions, although solids and gases can also be studied using this technique. Liquids are typically contained in transparent vessels made of materials like silica, glass, or plastic, known as cells or cuvettes. Gases can be enclosed in similar cells that are sealed to ensure that they are gas-tight.

In general, UV-Vis spectroscopy measurements are performed on biomaterial compounds that exhibit absorption in the UV-Vis range due to the presence of chromophores and auxochromes in their structure. These compounds undergo electronic transitions from the ground state to the excited state, leading to absorption in the UV-Vis range.

1.11.3 Sample Characteristics

Beer–Lambert's law states that the absorbance of a substance in a solution is directly proportional to its concentration, which makes absorption spectroscopy suitable for quantitative analysis. This principle establishes absorption spectroscopy as a highly effective analytical tool for characterizing and evaluating a wide range of materials, including dental biomaterials and dental composites.[41]

1.11.4 Advantages of Technique

UV-Vis spectroscopy is employed for both qualitative and quantitative analyses due to its notable advantages in sensitivity, precision in measurement, and selectivity in determinations. The sensitivity of spectrophotometric methods is quantitatively assessed using the molar absorptivity (ε), which correlates with the λ max of the solution being tested. Methods with ε values exceeding 10,000 $dm^3\ mol^{-1}\ cm^{-1}$ are considered highly sensitive, whereas those below 1000 $dm^3\ mol^{-1}\ cm^{-1}$ are less so.

1.11.5 Limitations of Technique

The accuracy of measurements depends on the concentration range and the quality of the instrumentation used, influenced by the efficiency of absorption and the specificity of reagents that produce a coloured product with the substance being analysed. Organic compounds containing chromophoric groups capable of undergoing electronic transitions from ground to excited states absorb radiation in the UV-Vis range. These chromophores are typically aromatic systems or groups with multiple bonds (e.g., >C=C<, >C=O, –N=N–, $-NO_2$).

1.11.6 Applications

UV-Vis spectroscopy has been employed in various applications within dental materials research. It has been used to analyse fillers used in dental adhesive resins and to evaluate the potential risks associated with chemical irritation, allergic reactions, and oral hypersensitivity due to polymer elution.[42] Furthermore, UV-Vis spectroscopy has been instrumental in assessing new co-initiators for their efficacy in radical photo-polymerization of dental polymers.[40] Additionally, this technique has been utilized to characterize the colour, spectral reflectance, and thickness of oxide layers and resins, as well as to evaluate the transparency of ceramics.[43] Moreover, UV-Vis spectroscopy has been utilized to assess the degradation rate of CH-based scaffolds, which are crucial in endodontic and periodontal regeneration materials.[43,44]

1.12 X-RAY SPECTROSCOPY[45–47]

1.12.1 Basic Working Principle

X-rays belong to the electromagnetic wave spectrum with wavelengths ranging from 10^{-9} to 10^{-12} meters. They are produced in an X-ray tube when electrons emitted by the cathode bombard the anode (anti-cathode) and are accelerated to high speeds in an electric field. The X-ray spectrum emitted from the anode material contains characteristic lines corresponding to specific electron transitions of the element.

X-ray analysis involves separating these lines for detailed examination. X-rays are detected using either energy-dispersive spectrometers or wavelength-dispersive spectrometers (WDSs). Analysis can be performed point-by-point or through mapping techniques. Quantitative elemental analysis is conducted by comparing the obtained data with standard reference materials.

Various methods of X-ray spectroscopy are employed in testing dental materials, including X-ray absorption spectroscopy (XAS), X-ray emission spectroscopy (XES), X-ray fluorescence spectroscopy (XRF), X-ray photoelectron spectroscopy (XPS), and X-ray diffraction (XRD). XAS investigates the relationship between X-ray intensities before and after passing through a sample of defined thickness. XES studies the relaxation products of photon or electron interactions with the sample, focusing on the emission of X-ray photons in dipole transitions between electronic states. XRF is an analytical method that measures the emission of characteristic secondary or fluorescent X-rays from a sample excited by high-energy radiation. A relatively recent technique, total-reflection X-ray fluorescence (TXRF), leverages the total reflection effect. This technique involves irradiating a sample, typically placed on a quartz support, with primary radiation from an X-ray tube at an angle smaller than the critical angle. XPS analyses the kinetic energy of electrons emitted from the top layers (1–10 nm) of the material upon exposure to X-rays. XPS provides information about the elements present in ultra-thin and surface layers of the material, as well as their chemical states and quantities.

XRD measures the intensity and scattering angles of X-rays that have passed through or been reflected from a material, maintaining the same energy as the incident radiation. Numerous databases exist containing XRD spectra, facilitating rapid phase identification for a wide range of crystalline samples.

1.12.2 Type of Samples Tested

In the realm of dental materials testing, XRF and XRD are the primary X-ray techniques employed. XRF is utilized for the analysis of metals, ceramics, and glass, accommodating samples of various shapes such as powders, pastes, solids, or liquids. On the other hand, TXRF enables the analysis of solutions, thin layers, solids, and different surfaces.

XRD primarily involves the examination of solids, necessitating a preparation with a flat and smooth surface. The sample is typically prepared by filling the window of an aluminium holder with a powder sample, followed by gently pressing a glass plate onto the surface of the preparation.[46,48]

1.12.3 Sample Characteristics

X-ray methods enable the analysis of the elemental composition in a localized area near or on the surface of a test sample. For instance, changes in microstructure and elemental composition of various restorative dental biomaterials under high temperatures can be observed and quantified using X-ray spectroscopy to detect trace elements. The capability to differentiate dental materials through elemental analysis has significantly enhanced the identification process.[49,50]

1.12.4 Advantages of Technique

XRF is employed for both qualitative and quantitative analysis, offering excellent selectivity and a low quantification limit. This method is particularly valuable for surface analysis. In trace analysis, XRF can detect substances at concentrations as low as 10^{-9} grams in a sample weighing approximately 0.1 grams. An additional advantage is that XRF is non-destructive.

TXRF extends the capabilities of XRF by detecting trace amounts of elements ranging from 10^{-7} to 10^{-12} grams in a 1 gram sample. TXRF can use very small samples, approximately 5×10^{-2} cm^3 in size.[51]

1.12.5 Limitations of Technique

In XRF, special crystals are necessary instead of diffraction gratings, and the method is suitable for elements with low excitation potentials but may be affected by matrix effects. On the other hand, XRD is specifically designed to characterize crystalline materials.

1.12.6 Applications

XRF is utilized for analysing material composition, measuring coating thickness, and identifying elements within dental biomaterials such as implant alloys and ceramics.[51] On the other hand, XRD is employed to gather details about the crystalline structure of solid materials, including dental ceramics.[51]

1.13 MASS SPECTROMETRY[52–55]

1.13.1 Basic Working Principle

MS is an analytical technique employed for structural confirmation of organic compounds and for qualitative and quantitative analysis of specific compounds within mixtures. It enables detection of substances even in trace amounts (as low as femtograms) in complex chemical mixtures. Regardless of their design, all spectrometers operate by ionizing the molecules of the substance being tested, which are then accelerated in an electric field under vacuum conditions. The resulting heterogeneous flux of positive or negative ions is separated into various components based on their mass-to-charge ratio (m/z). Typically, the molecular weight of the compound or its fragments can be deduced from the m/z ratio of ions detected.

In many mass spectrometers, ionization methods are selected such that most ions carry a charge (z) of 1. This simplifies interpretation of the spectrum, where m/z generally corresponds directly to the molecular weight of the ion. Ionization of samples in MS can be achieved through various methods, including electron ionization (EI), electrospray ionization (ESI), fast-atom bombardment (FAB), secondary ion mass spectrometry (SIMS), laser desorption (LD), matrix-assisted laser desorption/ionization (MALDI), and inductively coupled plasma mass spectrometry (ICP-MS).

1.13.2 Type of Samples Tested

Samples suitable for SIMS testing include metals, ceramics, organic and biological materials, polymers, biomaterials, and composites. They need to be compatible with high vacuum conditions, possess a flat surface with minimal topography, and should be free of any loose particles that could potentially enter and harm the mass analyser.

1.13.3 Sample Characteristics

Analysis of the data obtained from MS facilitates the identification of chemical compounds and enables precise determination of their composition in complex mixtures, particularly those with high molecular weights. This analytical tool is pivotal in research involving dental materials and polymer chemistry. For instance, it can elucidate the organic composition and eluates of three resin-based pulp coatings in relation to their specified uses and safety data sheets. Moreover, MS plays a crucial role in assessing the molecular toxicology of substances released from dental biomaterials.[56]

1.13.4 Advantages of Technique

Time of Flight Secondary Ion Mass Spectrometry (ToF-SIMS) is a qualitative technique employed to analyse surface characteristics, offering insights into molecular compounds, typically fragments of larger organic macromolecules found on the sample's outermost surface.

Recognized for its high sensitivity, SIMS combines high mass resolution and the capability to quantify elements, albeit restricted to analysing a fixed number of predefined ions simultaneously (up to seven ions). Given its sensitivity, it requires the use of high purity solvents only.

1.13.5 Disadvantages of Technique

SIMS is limited in optical capabilities, facing challenges in collecting positive or negative ion data, and depending on the sample type, analysis can range from 30 minutes to 5 hours. Post-testing, the sample surface may exhibit damage.

1.13.6 Applications

Studies documented so far have utilized SIMS to investigate the composition and chemical interactions of dental resins, ceramics, and composite materials.[51,57]

In a narrative review, Kaczmarek et al. provided illustrative diagrams depicting the spectrum of outcomes observed during testing of various dental biomaterials with various techniques discussed above in the chapter (Figure 1.4).[51]

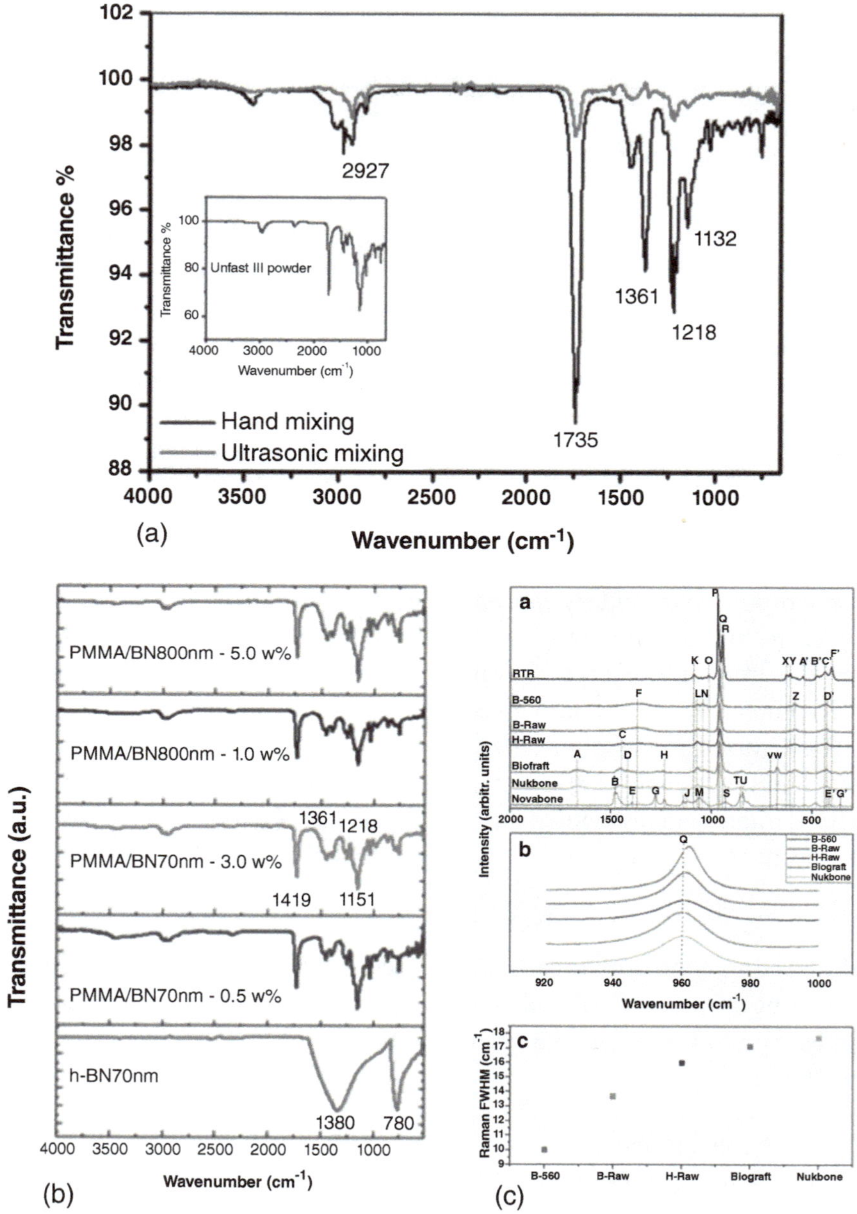

FIGURE 1.4 Analysis of various dental biomaterials using different spectroscopic methods: (a and b) FT-IR data for the different tested samples; (c) Raman spectra of the various bone grafts tested;

(Continued)

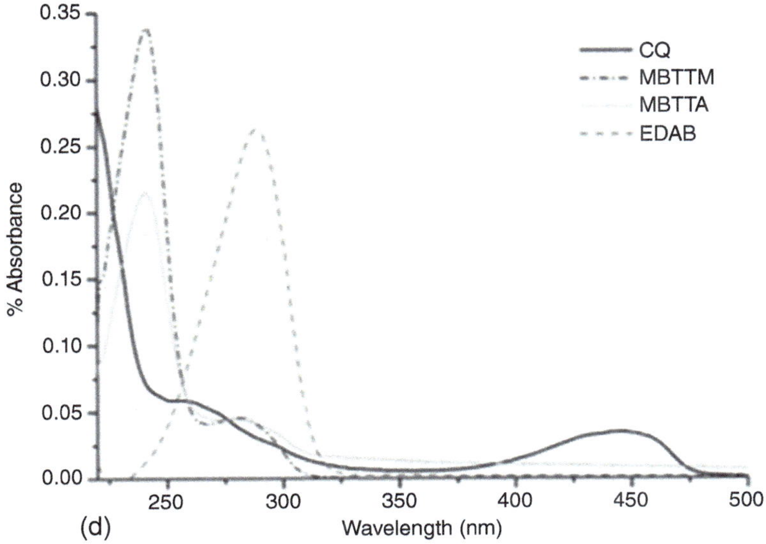

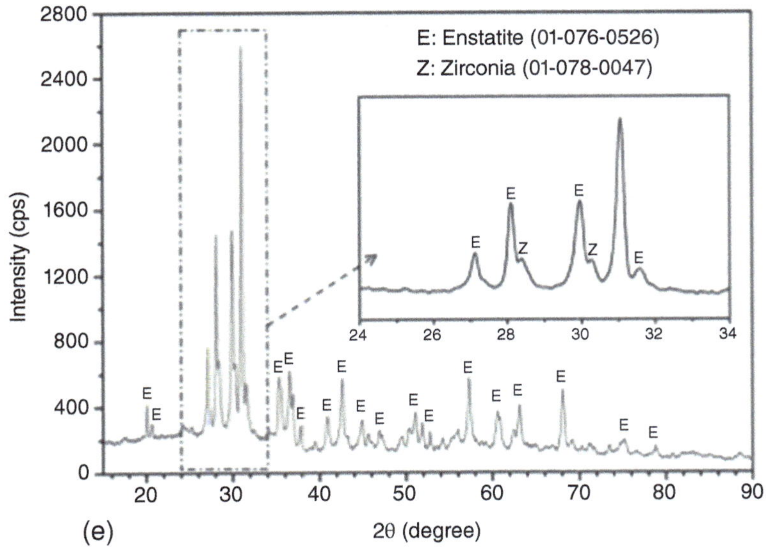

FIGURE 1.4 (Continued) (d) UV-Vis spectra; and (e) XRD patterns of the tested samples.

1.14 CONCLUSION AND FUTURE ASPECTS

The chapter explores various spectroscopic methods used to analyse surfaces and their applications in studying advanced dental biomaterials, encompassing ceramics, steel, glasses, cements, composites, and resins. These methods include IR (FT-IR, Raman), UV-Vis spectroscopy, X-ray spectroscopy (XAS, XES, XRF, XRD), and MS (ToF-SIMS). Each spectroscopic technique offers specific strengths and applications, requiring careful selection based on the material being studied. The effectiveness of each method depends on its ability to characterize different aspects of the material's composition and structure.

Researchers, clinicians, and dental practitioners need a comprehensive understanding of these techniques' applications in dental materials research, recognizing that no single method can provide complete information across all parameters. Therefore, researchers typically employ multiple complementary methods to validate their findings, covering elemental composition, chemical properties, and structural characteristics of emerging dental biomaterials. This multi-method approach is essential in dental materials research, especially where standardized reference biomaterials are lacking for various dental applications.

Advancements in these techniques include their integration with other methods, such as combining atomic force microscopy with IR to create AFM-IR for nanoscale chemical analysis, which is progressing. Additionally, there is anticipation regarding the development of alternative methodologies, such as electrochemical or Raman microscopy for label-free biosensing. Future developments will focus on improving equipment and computational methods to introduce new functionalities that enhance data quality, accuracy, and interpretability. It remains critical to address matrix effects, particularly significant in thicker dental biomaterial preparations, to ensure precise analysis.

In the characterization of dental biomaterials, mechanical properties were predominantly assessed for restorative materials, while biological properties were emphasized for dental implants and endodontic materials. Physical properties received particular attention in the study of bioceramics. Currently, there is a growing emphasis on restorative materials that offer enhanced aesthetics, such as resin composites, adhesive resin composites (luting cements), zirconia, and other ceramics, to meet clinical demands. Increased research focus on bioceramics reflects their regenerative potential, contributing to expanded laboratory and animal studies in this field.

While significant strides have been made in employing biomolecules within suitable scaffolds for effective stem cell transplantation in addressing diseases resistant to conventional treatments, ongoing efforts are required to identify these molecules and comprehend their interactions with extracellular matrices (ECMs) and differentiated stem cells (DSCs) to optimize therapeutic outcomes. There remains a gap in studies utilizing spectroscopic techniques to analyse the interaction between biological cells, tissue matrices, and autologous biomaterials such as platelet-rich fibrin, platelet-rich plasma derived from patients, and stem cells for tissue engineering.

This chapter provides invaluable insights for emerging researchers, guiding future research directions and project planning in the fields of biomaterials for tissue engineering and the regeneration of soft and hard tissues.

REFERENCES

1. Williams DF. *The Williams Dictionary of Biomaterials.* Liverpool University Press, Liverpool, UK 1999.
2. Cohen J. Biomaterials in orthopedic surgery. *Am J Surg* 1967;114:31–41.
3. Reddy LRM, Reddy MO, Thaidun M, Reddy PG. Biomaterial and its applications. *J Inf Comput Sci* 2020;10:90–4.
4. Chelu M, Musuc AM. Advanced Biomedical Applications of Multifunctional Natural and Synthetic Biomaterials. *Processes* 2023;11:2696.
5. Agrawal R, Kumar A, Mohammed MKA, Singh S. *Biomaterial types, properties, medical applications, and other factors: a recent review. Journal of Zhejiang University-SCIENCE A (Applied Physics & Engineering)* Zhejiang University Press 2023:1–16.
6. Zhang YS, Yao J. Imaging Biomaterial-Tissue Interactions. *Trends Biotechnol* 2018;36:403–14.
7. Kaczmarek K, Leniart A, Lapinska B, Skrzypek S, Lukomska-Szymanska M. Selected Spectroscopic Techniques for Surface Analysis of Dental Materials: A Narrative Review. *Materials (Basel)* 2021;14:2624.
8. Sharma RK. Various Spectroscopic Techniques. In *Environmental Pollution: Monitoring Modelling and Control*; Studium Press, LLC: Houston, TX, USA 2017:181–206.
9. Kafle BP. Infrared (IR) spectroscopy. In *Chemical Analysis and Material Characterization by Spectrophotometry*; Elsevier: New York, NY, USA 2020:199–243.
10. Brader ML. UV-absorbance, fluorescence and FT-IR spectroscopy in biopharmaceutical development. In *Biophysical Characterization of Proteins in Developing Biopharmaceuticals*; Elsevier: New York, NY, USA 2020:97–121.
11. Bell SEJ, Xu Y. Infrared spectroscopy|Industrial applications. In Encyclopedia of Analytical Science; Elsevier: New York, NY, USA 2019:124–33.

12. Kowalczuk D, Pitucha M. Application of FTIR Method for the Assessment of Immobilization of Active Substances in the Matrix of Biomedical Materials. *Materials* 2019;12:2972.
13. Dutta A. Fourier Transform Infrared Spectroscopy. In *Spectroscopic Methods for Nanomaterials Characterization*; Elsevier: New York, NY, USA 2017;2:73–93.
14. Alotaibi J, Saji S, Swain MV. FTIR characterization of the setting reaction of biodentine. *Dent Mater* 2018;34:1645–51.
15. Puspita S, Sunarintyas S, Mulyawati E, Anwar C, Sukirno, Soesatyo MHNE. Molecular weight determination and structure identification of Bombyx mori L. Fibroin as material in dentistry. In *AIP Conference Proceedings*; American Institute of Physics Inc: College, MA, USA 2020;2260:40018.
16. Rafeek AD, Choi G, Evans LA. Morphological, spectroscopic and crystallographic studies of calcium phosphate bioceramic powders. *J Aust Ceram Soc* 2018;54:161–8.
17. Puppin-Rontani J, Fugolin APP, Costa AR, Correr-Sobrinho L, Pfeifer CS. In vitro performance of 2-step, total etch adhesives modified by thiourethane additives. *Int J Adhes Adhes* 2020;103:102688.
18. Shim JS, Lee SY, Song SY, Jha N, Ryu JJ. Polymerization efficiency of dental dual-cured resin cement light-cured at various times after the initiation of chemical activation. *Int J Polym Mater Polym Biomater* 2020;69:622–8.
19. Fugolin AP, Lewis S, Logan MG, Ferracane JL, Pfeifer CS. Methacrylamide-methacrylate hybrid monomers for dental applications. *Dent Mater* 2020;36:1028–37.
20. Fugolin AP, Dobson A, Ferracane JL, Pfeifer CS. Effect of residual solvent on performance of acrylamide-containing dental materials. *Dent Mater* 2019;35:1378–87.
21. Fugolin APP, Navarro O, Logan MG, Huynh V, França CM, Ferracane JL, et al. Synthesis of di- and triacrylamides with tertiary amine cores and their evaluation as monomers in dental adhesive interfaces. *Acta Biomater* 2020;115 148–59.
22. Alania Y, dos Reis MC, Nam JW, Phansalkar RS, McAlpine J, Chen SN, et al. A dynamic mechanical method to assess bulk viscoelastic behavior of the dentin extracellular matrix. *Dent Mater* 2020;36:1536–43.
23. Zhang P, Zhao XM. Synthesis, crystal structure and bioactivity evaluation of a heterocyclic compound. *Jiegou Huaxue* 2020;39:1892–7.
24. Seredin PV, Uspenskaya OA, Goloshchapov DL, Ippolitov IY, Vongsvivut J, Ippolitov YA. Organic-mineral interaction between biomimetic materials and hard dental tissues. *Sovrem Tehnol V Med* 2020;12:43–51.
25. Lach S, Jurczak P, Karska N, Kubiś A, Szymańska A, Rodziewicz-Motowidło S. Spectroscopic Methods Used in Implant Material Studies. *Molecules* 2020;25:579.
26. Jing X, Xie B, Li X, Dai Y, Nie L, Li C. Peptide decorated demineralized dentin matrix with enhanced bioactivity, osteogenic differentiation via carboxymethyl chitosan. *Dent Mater* 2021;37:19–29.

27. Fu D, Lu Y, Gao S, Peng Y, Duan H. Chemical Property and Antibacterial Activity of Metronidazole-decorated Ti through Adhesive Dopamine. *J Wuhan Univ Technol Mater Sci Ed* 2019;34:968–72.
28. Kafle BP. Raman spectroscopy. In *Chemical Analysis and Material Characterization by Spectrophotometry*; Elsevier: New York, NY, USA 2020:245–68.
29. Xu Z, He Z, Song Y, Fu X, Rommel M, Luo X, et al. Topic review: Application of raman spectroscopy characterization in micro/nano-machining. *Micromachines* 2018;9:361.
30. Jones RR, Hooper DC, Zhang L, Wolverson D, Valev VK. Raman Techniques: Fundamentals and Frontiers. *Nanoscale Res Lett* 2019;14:1–34.
31. Tzanakakis E, Kontonasaki E, Voyiatzis G, Andrikopoulos K, Tzoutzas I. Surface characterization of monolithic zirconia submitted to different surface treatments applying optical interferometry and raman spectrometry. *Dent Mater J* 2020;39:111–7.
32. Malhotra R, Han YM, Morin JLP, Luong-Van EK, Chew RJJ, Castro Neto AH, et al. Inhibiting Corrosion of Biomedical-Grade Ti-6Al-4V Alloys with Graphene Nanocoating. *J Dent Res* 2020;99:285–92.
33. Khan AS, Khalid H, Sarfraz Z, Khan M, Iqbal J, Muhammad N, et al. Vibrational spectroscopy of selective dental restorative materials. *Appl Spectrosc Rev* 2017;52:507–40.
34. Gomes de Araújo-Neto V, Sebold M, Fernandes de Castro E, Feitosa VP, Giannini M. Evaluation of physico-mechanical properties and filler particles characterization of conventional, bulk-fill, and bioactive resin-based composites. *J Mech Behav Biomed Mater* 2021;115:104288.
35. Zubieta-Otero LF, Londoño-Restrepo SM, Lopez-Chavez G, Hernandez-Becerra E, Rodriguez-Garcia ME. Comparative study of physicochemical properties of bio-hydroxyapatite with commercial samples. *Mater Chem Phys* 2021;259: 124201.
36. Lubas M. Au interface effect on Ti-dental porcelain bond strength investigated by spectroscopic methods and mechanical tests. *J Mol Struct* 2020;1208:127870.
37. Ubaldini ALM, Pascotto RC, Sato F, Soares VO, Baesso ML. Mechanical and Chemical Changes in the Adhesive-Dentin Interface after Remineralization. *J Adhes Dent* 2020;22:297–309.
38. Lubas M, Przerada I, Zawada A, Jasinski JJ, Jelen P. Spectroscopic and microstructural investigation of novel Ti-10Zr-45S5 bioglass composite for dental applications. *J Mol Struct* 2020;1221:128545.
39. Kafle BP. Application of UV-VIS spectrophotometry for chemical analysis. In *Chemical Analysis and Material Characterization by Spectrophotometry*; Elsevier: New York, NY, USA 2020:79–145.
40. Keul C, Seidl J, Güth JF, Liebermann A. Impact of fabrication procedures on residual monomer elution of conventional polymethyl methacrylate (PMMA)-a measurement approach by UV/Vis spectrophotometry. *Clin Oral Investig* 2020;24:4519–30.

41. Omidi M, Fatehinya A, Farahani M, Akbari Z, Shahmoradi S, Yazdian F, et al. Characterization of biomaterials. In *Biomaterials for Oral and Dental Tissue Engineering*; Elsevier: New York, NY, USA 2017:97–115.
42. Stürmer M, Garcia IM, Souza VS, Visioli F, Scholten JD, Samuel SMW, et al. Titanium dioxide nanotubes with triazine-methacrylate monomer to improve physicochemical and biological properties of adhesives. *Dent Mater* 2021;37:223–35.
43. Michailova M, Elsayed A, Fabel G, Edelhoff D, Zylla IM, Stawarczyk B. Comparison between novel strength-gradient and color-gradient multilayered zirconia using conventional and high-speed sintering. *J Mech Behav Biomed Mater* 2020;111:103977.
44. Qasim SB, Najeeb S, Delaine-Smith RM, Rawlinson A, Ur Rehman I. Potential of electrospun chitosan fibers as a surface layer in functionally graded GTR membrane for periodontal regeneration. *Dent Mater* 2017;33:71–83.
45. Fracchia M, Ghigna P, Vertova A, Rondinini S, Minguzzi A. Time-Resolved X-ray Absorption Spectroscopy in (Photo)Electrochemistry. *Surfaces* 2018;1:11.
46. Khan H, Yerramilli AS, D'Oliveira A, Alford TL, Boffito DC, Patience GS. Experimental methods in chemical engineering: X-ray diffraction spectroscopy-XRD. *Can J Chem Eng* 2020;98:1255–66.
47. Yahia LH, Mireles LK. X-ray photoelectron spectroscopy (XPS) and time-of-flight secondary ion mass spectrometry (ToF SIMS). In *Characterization of Polymeric Biomaterials*; Elsevier: New York, NY, USA 2017:83–97.
48. Bunaciu AA, Udri Stioiu EG, Aboul-Enein HY. X-ray Diffraction: Instrumentation and Applications. *Crit Rev Anal Chem* 2015;45:289–99.
49. Luo Q. Electron Microscopy and Spectroscopy in the Analysis of Friction andWear Mechanisms. *Lubricants* 2018;6:58.
50. Çarıkçıo Glu B, Misilli T, Deniz Y, Aktas Ç. Effects of high temperature on dental restorative materials for forensic purposes. *Forensic Sci Med Pathol* 2021;17:78–86.
51. Kaczmarek K, Leniart A, Lapinska B, Skrzypek S, Lukomska-Szymanska M. Selected Spectroscopic Techniques for Surface Analysis of Dental Materials: A Narrative Review. *Materials* 2021;14:2624.
52. Gosetti F, Marengo E. Mass spectrometry|Selected ion monitoring. In *Encyclopedia of Analytical Science*; Elsevier: New York, NY, USA 2019:500–10.
53. Lermyte F. Modern Mass Spectrometry and Advanced Fragmentation Methods. In *New Developments in Mass Spectrometry*; Royal Society of Chemistry: Cambridge, UK 2021:1–14.
54. Schaepe K, Jungnickel H, Heinrich T, Tentschert J, Luch A, Unger WES. Secondary ion mass spectrometry. In *Characterization of Nanoparticles*; Elsevier: New York, NY, USA 2020:481–509.
55. Paital B. Mass Spectrophotometry: An Advanced Technique in Biomedical Sciences. *Adv Tech Biol Med* 2015 4:1–8.

56. Bakopoulou A, Papadopoulos T, Garefis P. Molecular toxicology of substances released from resin-based dental restorative materials. *Int J Mol Sci* 2009;10:3861–99.
57. França R, Samani TD, Bayade G, Yahia L, Sacher E. Nanoscale surface characterization of biphasic calcium phosphate, with comparisons to calcium hydroxyapatite and tricalcium phosphate bioceramics. *J Colloid Interface Sci* 2014;420:182–8.

CHAPTER 2

Fourier Transform Infrared Spectroscopy for Analysis of Dental Materials

Current Trends

Jitendra Sharan*, Anisha Mishra*, Manoj Pratap Singh, and Veena Koul

2.1 INTRODUCTION

Dental materials are specifically fabricated materials intended to be used in dentistry for various purposes. These can be in the form of replacement of missing hard and soft tissue structures and enhancement of the facial aesthetics, form, and functions,[1–4] thus enhancing the health and functioning of the stomatognathic system. In the last few decades, with the rapid evolution of the dental materials, a wide variety of dental materials are currently available for oral application (Figure 2.1).[5–7]

The rapid evolution in the field of dental biomaterials is secondary to the better understanding of the dental hard and soft tissue structures (such as mechanical, physical, thermal, and rheological) and advancement in the research and material technology (Figure 2.2).[4,8–10] For dental materials to function optimally in the oral environment (which is one of the most challenging biological environments), the close matching

* Authors contributed equally

DOI: 10.1201/9781003506805-2

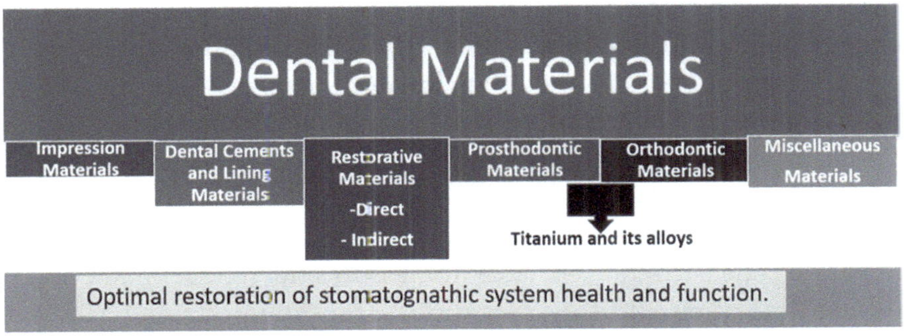

FIGURE 2.1 Dental materials system currently available for application in oral health science.

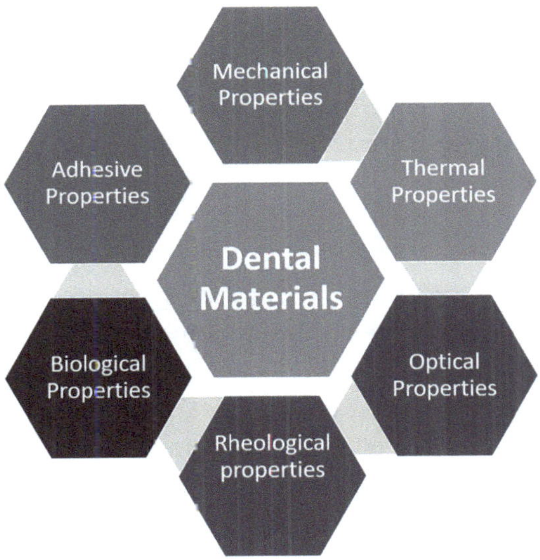

FIGURE 2.2 Various important properties of dental materials.

of the various properties of it to that of the biological tissues is of paramount significance.[4,6] Over decades, various testing methods have been designed and developed to assess mechanical, surface, and biological properties to make sure that the dental materials perform close to ideal in vitro and in vivo conditions (in the oral cavity of the patients). Tests such as Fourier transform infrared (FTIR) spectroscopy, scanning electron microscopy (SEM), atomic force microscopy (AFM), X-ray diffraction (XRD), and X-ray photo spectrometry (XPS) are sophisticated and state-of-the-art tools for the study of surface topography as well as quantitative and qualitative assessment of chemical species present in the dental materials.[11–20] These tools also enable the researchers to comprehensively study the surface phenomena and chemical structure of the dental materials. Furthermore, they allow for advanced experimental research, which includes comprehensive data acquisition, data processing, and its interpretation.

FTIR is one such tool available. It measures the interaction between infrared radiation and the spectrum of absorption of solid, gas, and liquid.[21,22] In this, the infrared radiation is transmitted through a sample in infrared spectroscopy. The spectrum which is obtained provides the details of molecular fingerprint by displaying the molecule's absorption and transmission. FTIR spectroscopy is commonly used in organic synthesis, pharmaceutical sciences, and polymer science for various aspects of molecular validation and confirmation.[23,24] They can act as a powerful tool to explore the mechanism of chemical reactions and the detection of unstable chemical compounds when used along with chromatography.

2.2 PHYSICS AND WORKING PRINCIPAL OF FTIR[25–27]

FTIR stands for Fourier transform infrared spectroscopy. Infrared spectroscopy provides knowledge on the vibrational and rotational modes of motion of a molecule and is therefore an important technique for identification and characterization of functional groups and bonds between atoms. The infrared spectra of an organic compound present a unique fingerprint, which is quickly differentiated from the absorption patterns of all other compounds. The spectra of an organic compound can be used to quantify some components of an unknown mixture. It can be applied for the analysis of *solids*, *liquids*, and *gaseous* samples.

2.2.1 Why Use FTIR Spectroscopy (FTIR Analysis)

FTIR is non-destructive and can identify unknown materials. It can measure the quantity of components in a mixture and can be configured to measure solids, liquids, and gases. Measurements made by FTIR are complete in just a few seconds, since all frequencies are measured at once. Sensitivity is significantly improved with lower noise levels and ultrasensitive detectors. Heightened sensitivity can detect even the smallest contaminants, e.g., gas standards.

2.2.2 Comparison of Raman and IR Spectroscopy

2.2.2.1 Mutual Exclusion Principle

In a molecule with a centre of symmetry, it is observed that vibrations that are **Raman active** are **IR inactive** and **vice versa**; this is known as the principle of mutual exclusion (e.g., as in CO_2). Some molecules have different elements of symmetry as they have certain bands that may be active in IR, Raman, both, or neither. A complex molecule that has no symmetry except for a characteristics element will have all of the normal modes active in both IR and Raman. However, this does not mean that they can be monitored. In both types, the neighbouring strong bands may obscure weak bands, while others may be intrinsically too weak to be observed even if they are theoretically "allowed".

In general, the strong bands in the IR spectrum of a compound correspond to weak bands in the Raman and vice versa. The electrical characteristic of the vibration is the cause of such complimentary nature.

If a bond is strongly polarized, a little alteration in its length which occurs during a vibration will have only a little additional effect on polarization. Vibrations involving polar bonds O-H, N-O, C-O, and so on have comparatively weak Raman scatterers. Such polarized bonds, however, carry their charges during the vibrational motion (unless neutralized by symmetry factors), which results in a large net dipole moment change and produces strong IR absorption band. Conversely, relatively neutral bonds (C-C, C-H, C=C) suffer large changes in polarizability during a vibration, though this is less easy to visualize. Although the dipole moment is not affected in a similar way, vibrations that broadly involve this type of bond are strong Raman scatterers but show weak IR signals.

2.2.2.2 Mutual Exclusion Principle as Seen in CO_2

Molecules that have an inversion centre will have none of the normal modes of vibrations; therefore, they are neither Raman active nor IR active. This is known as the **mutual exclusion principle**. A simple molecule that obeys this principle is CO_2. Carbon dioxide has an inversion centre or centre of symmetry. Figure 2.3 depicts its normal modes of vibrations. The Raman and IR-active modes are shown below each type of vibration.

2.2.2.3 Selection Rules for Vibrations

Vibrational modes can be comprised of two modes: (i) stretching (symmetric and asymmetric) and (ii) bending vibrations (in-plane and out-of-plane). The selection rule states that the vibrations can occur in

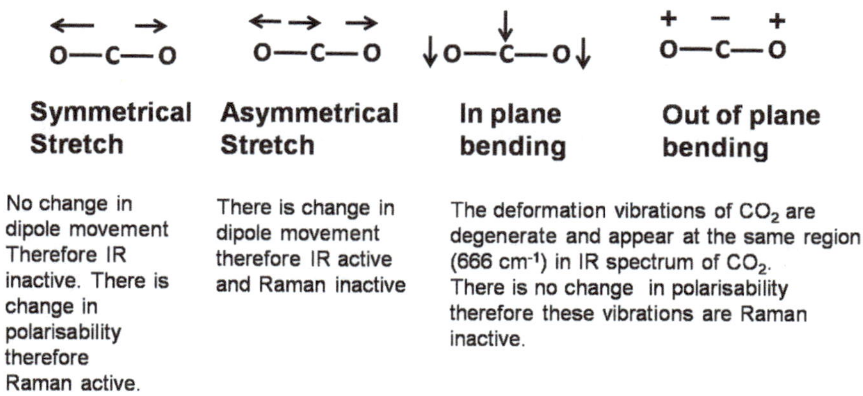

FIGURE 2.3 Different types of vibrations in a molecule.

only **IR-active** compounds, if the change in molecular dipole moment is observed during the vibration. If a diatomic molecule is having the same atoms, it cannot be excited to vibrate because no dipole moment exists between them. In contrast to that, molecules with a variety of types of atoms can interact with incident radiation even if a dipole moment does not exist in the beginning; it can be induced due to anti-symmetric displacement of the centre of charge (e.g., CO_2).

A molecule of *n* atoms has 3N degrees of freedom. Three degrees are rotational and translational in a non-linear molecule, and the remaining are associated with fundamental vibrations (normal modes). Two and three degrees are **rotational** and **translational**, respectively, in a linear molecule. The number of fundamental vibrations in a linear molecule and in a non-linear molecule is 3N-5 and 3N-6, respectively. H_2O is a good example of a non-linear molecule, and it has three fundamental vibrations ($3 \times 3 - 6$). The CO_2 is a good example of a linear molecule, and it has four fundamental vibrations ($3 \times 3 - 5$).

i. **Degenerated vibrations**: Symmetrically equivalent vibrations have same wave number, e.g., CO_2 bending vibrations

ii. **Coupled vibrations**: Bands with slightly different positions can occur due to the network of molecules in a structure

iii. **Overtone bands**: Due to transitions of $2h\nu$, $3h\nu$, etc., lower intensity is observed than the fundamental vibration band. Band position is seen at a slightly smaller frequency than the multitude of the fundamental vibrational frequency.

iv. **Combination vibrations**: This is the sum or difference of two vibrations; they can be supportive to determine the speciation of a molecule in a sample.

2.2.3 Description of FTIR Spectrometer

FTIR spectrometry was developed to break the boundaries related to dispersive instruments. The main problem was the slow scanning speed. A method for computing all of the infrared frequencies at the same time, rather than individually, was required. To overcome this problem, an interferometer was developed, which is a very simple optical device. This optical device generates a unique type of signal that has all of the IR frequencies "encoded" into it. The signal can be calculated very quickly, generally on the order of one second or so. Consequently, the time element per sample is decreased to a matter of a few seconds instead of several minutes. Most interferometers utilize a beamsplitter which takes the incoming IR beam and splits it into two optical beams. One beam reflects off a flat mirror which is fixed in place. The other beam reflects off of a flat mirror which is based on a mechanism which permits this mirror to shift a very small distance (typically a few millimetres) away from the beamsplitter. The two beams reflect off of their individual mirrors and are reunited when they assemble back at the beamsplitter. As the path length of one beam is fixed and the other is at a constant rate varying as its mirror shifts, the signal that departs the interferometer is the product of these two beams "interfering" with each other. The resultant signal is known as an interferogram which has the unique characteristic that every data point (a function of the shifting mirror position) that makes up the signal has information about every IR frequency that comes from the source. The interferogram is then calculated, and all frequencies are measured concurrently. Therefore, employing the interferometer results in exceptionally fast measurements.

The analyst needs a frequency spectrum (a graph plot of the intensity vs. individual frequency) for identification. The interpretation of measured interferogram signal cannot be directly possible. For achieving this, "decoding" the individual frequencies is necessary. This can be achieved through a recognized mathematical technique known as the Fourier transformation. This transformation is carried out by the computer, and the analyst obtains the desired spectral information for analysis.

2.2.3.1 The Sample Analysis Process

The normal instrumental process is as follows (Figure 2.4):

1. **The Source**: IR energy is emitted from a radiant black-body source. This beam goes by an opening which regulates the amount of energy accessible to the sample (and ultimately to the detector).
2. **The Interferometer**: The beam comes into the interferometer where the "spectral encoding" happens. The resultant interferogram signal then departs from the interferometer.
3. **The Sample**: The beam comes into the sample compartment where it is passed through or reflected/touched off of the surface of the sample, based on the type of analysis being achieved. This is the place where specific frequencies of energy, which are unique characteristics of the sample, are absorbed.
4. **The Detector**: The beam finally reaches the detector for final measurement. The detectors employed here are specifically planned to measure the special interferogram signal.
5. **The Computer**: The final measured signal comes out from detector in digital form and is passed on to the computer where the Fourier transformation occurs. The final IR spectrum is then provided to the analyst for interpretation and any further inference.

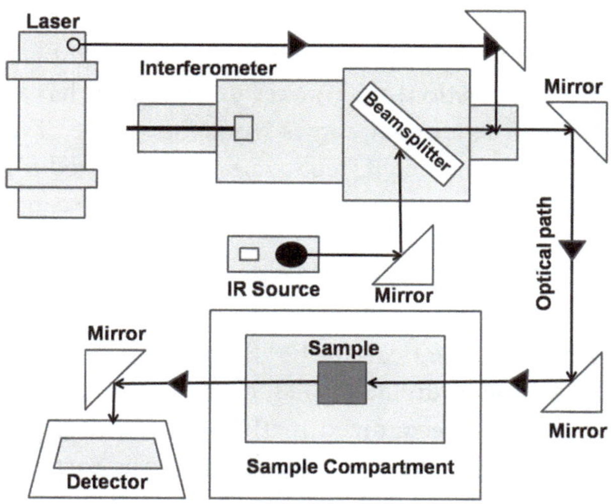

FIGURE 2.4 Sketch diagram of optical path of FTIR spectroscopy instrument.

FTIR can be employed to identify chemicals from paints, polymers, spills, coatings, contaminants, and drugs. FTIR is possibly the most powerful tool for distinguishing the different types of chemical bonds (functional groups). The wavelength of light absorbed is the unique characteristic of the chemical bond. FTIR spectra of pure compounds are normally so unique that they are exactly similar to the molecular "fingerprint". Generally, inorganic compounds are usually much simpler while organic compounds are very rich and have detailed spectra. Mostly for the common materials, the spectra of an unknown compound can be identified by comparing it to a library of known compounds. To identify fewer common materials, infrared will require to be united with mass spectrometry (MS), nuclear magnetic resonance (NMR), atomic emission spectroscopy (AES), XRD, and/or other techniques.

2.2.4 Description of FTIR Microscope

FTIR imaging is a complimentary imaging tool and is a very versatile analytical technique for spectra-chemical imaging. The advantage of chemical imaging in comparison to other analytical technologies is the capability to analyse the spatial distribution of different component materials in blends, granules, or finished dosage forms. Many researchers are interested in obtaining the molecular information from a specific area of a sample to study a picture of the distribution of molecules or functional groups. These chemicals, called infrared images, offer information, which is highly complementary to images acquired from techniques like visible light microscopy, SEM, and AFM. AFM and SEM have each a much superior spatial resolution than FTIR imaging, close to less than 1 nm. These techniques present the information about surface properties like morphology and topography and, in some cases, also provide elemental distribution, but they do not give information about molecular composition. FTIR spectroscopy is well established as a highly sensitive, fast technique for material characterization of a variety of samples for molecular species in a wide range of materials. By coupling the FTIR to a microscope accessory, measurements on small samples can be routinely carried out, down to a few micron areas. Applications include the identification of trace contaminants, the analysis of failure modes, and the characterization of production defects. FTIR takes advantage of the fact that the functional groups of every module generate a characteristic absorption or transmission spectral fingerprint that definitely identifies that chemical compound. FT-IR and FT-NIR (near infrared) are rapidly becoming the preferred compound testing technologies due to their speed, accuracy, and reliability.

While IR microscopy presents very high sensitivity and a broad spectral range for a small/specific area, FTIR measurements obtaining a microscopic chemical picture of a sample can be a cumbersome job depending on the data collection and sample size requirements. Unlike IR microscopes that use a single detector measuring a single point at a time, FTIR imaging systems have multi-element detectors, generating infrared images almost as fast as an optical microscope, which presents a visible image.

2.3 APPLICATIONS OF FTIR IN ORAL HEALTH SCIENCES

In dentistry, FTIR has a broad range of applications such as the analysis of polymerization kinetics, the structure of heterocyclic compounds for pulp revitalization, biomimetic and new-age materials replicating enamel and dentin complex, determination of intermolecular cross-linking of collagen and dentin, and the structural characterization of various dental materials.

Human enamel studies[28,29] using FTIR in the spectral range from 2000 to 900 cm^{-1} have shown that the absorption band related to the stretching vibrations has its peak at 1035 cm^{-1}. This peak was narrow and has weakly expressed "shoulder" in the short wavelength side. Amide I and II bands were clearly visible in carious and non-carious enamels, but Amide III absorption bands were remarkable for the carious enamel (Figure 2.5). Carious enamel is characterized by an increase in the number of stretching vibrations and deformation of N–C–O and C=O bonds, suggestive of decrease in crystallinity and texture indices along with decrease in the orientation of hydroxyapatite crystals. All these are clearly indicative of the destructive lesion in the organic matrix of the enamel.

Studies done on various zones of carious dentin have suggested that it could differentiate the sub-transparent zone of dentinal caries from the normal zone and transparent zone.[30] There was close similarity in the carbonate content between the sub-transparent zone and the normal zone. A-type carbonate-rich contents were seen in relation to both sub-transparent and transparent zones which could be suggestive of the fact that hydroxyapatite primarily occludes the dentinal tubules due to odontoblastic activity which later becomes whitlockite as the caries progresses due to de- or re-mineralization process.[31] FTIR absorption spectra have

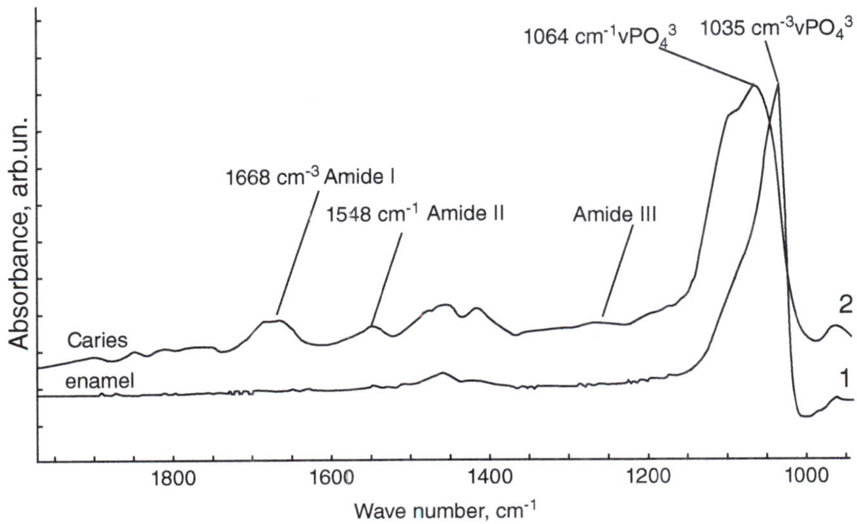

FIGURE 2.5 Prominent Amid III vibration bands of carious human enamel.[28]

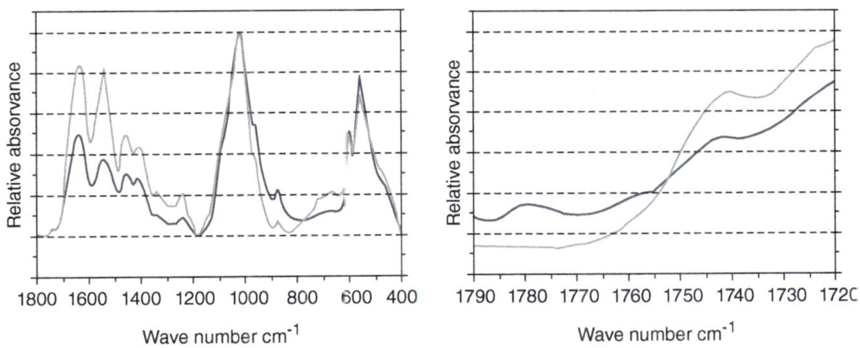

FIGURE 2.6 FTIR spectra of carious dentin showing the presence of carbonyl esters in the region of 1740 cm^{-1}.[30]

shown that the carious dentine contains ester groups, which are missing in the sound dentine, which is very crucial to differentiate the carious dental tissues (Figure 2.6).[29]

FTIR aids in the analysis of the polymerization kinetics of resins[32,33] and adhesives.[34–37] Delgado et al. studied various dental bonding systems for composition and polymerization kinetics using FTIR and suggested that this tool can be used precisely for estimation of the degree of conversion of the bonding system. The higher the degree of conversion, the better the bonding with the dental hard tissues.[38] ATR-FTIR is a useful method

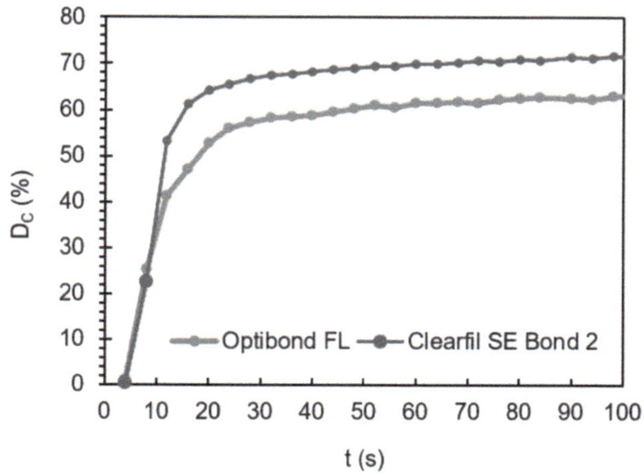

FIGURE 2.7 Degree of conversion of two bonding systems used in dentistry assessed by FTIR.[38]

to investigate relative levels of main components in bonding systems and their polymerization kinetics (Figure 2.7).

An experimental chitosan-based smart injectable hydrogel was assessed for release of anti-narcotic drugs naltrexone and disulfiram in phosphate buffer saline and different concentrations of ethanol over a 56-day period. It was noted that all formulations in phosphate buffer saline and only naltrexone infused in high molecular weight chitosan injectable hydrogel followed a non-Fickian type of drug release.[39]

FTIR analysis showed that when nanodiamonds with oxygen-containing functional groups were surface deposited on 3-D printed pro-caprolactone scaffolds, their hydrophilicity and degradation rate were increased, thereby increasing in vitro cell adhesion which could aid in regeneration potential.[40] When cellulose acetate scaffold was coated with a hydroxyapatite and graphene oxide in nanocomposite, FTIR could demonstrate characteristics of cellulose and PO_4 bands, thereby confirming the presence of CaP in the synthesized materials of the groups containing hydroxyapatite. Other analyses also proved the increase in cell adhesion and proliferation and its promising role in tissue engineering.[41]

FTIR revealed the presence of Si-O and CH chemical groups in strontium-doped bioglass-laden gelatine methacryloyl hydrogels which connotes a higher cytocompatibility, cell adhesion, and mineralization potential along with a relevant degradation profile supporting their use in vital pulp therapy procedures.[42]

FTIR analysis of chitosan nanoparticles combined with Salvodora persica showed the development of many active antimicrobial phytocompounds showing the synergistic effect in the form of mouth-rinsing solutions to control oropharyngeal microbes in viral secondary bacterial infections.[43]

Quaternary ammonium silane antimicrobial compound was coated to surgical sutures and analysed under FTIR. It peaked at 1490 cm^{-1} confirming the presence of Quaternary ammonium silane on suture's surfaces. Quaternary ammonium compounds tend to have strong antimicrobial properties in the clinical conditions and are widely used. The coating to the sutures with quaternary ammonium compound also enhances their physical and mechanical properties. Hence, it can be used a novel antimicrobial suture material.[44]

A novel carbon nanodot-stabilized silver nanoparticle-enriched calcium hydroxide endodontic sealer was introduced and tested to combat the persistent Faecalis bacteria in the root canals. Upon analysis by FTIR, it revealed absorption peaks to various functional groups showing high release of Ag+ ions, thereby possessing superior anti-bacterial properties.[45]

Elastomeric micelles were added to a high-strength glass ionomer cement, and the setting kinetics was analysed by FTIR. Although higher release of fluoride, aluminium, and phosphate ions was seen in conventional high-strength glass ionomer cement (GIC) powders, elastomeric micelles which were added to the high-strength glass ionomer cement exhibited satisfactory mechanical properties and cytocompatibility.[46]

Surface characterization of honey-based dental implants by FTIR showed the enhanced presence of antibacterial nanoparticles against bacterial strains of Streptococcus mutans and Streptococcus aureus.[47]

Recently, FTIR has been used for assessment of change in the surface functionality of grade V titanium alloy (Ti6Al4V) following treatment with 5M NaOH, APTES, and type I human collagen.[48] FTIR evaluation has suggested that there was attachment of hydroxyl group (–OH) after alkali treatment, to which APTES molecules got attached, and finally to this type I human collagen was bio-conjugated through EDC-NHS coupling reaction (Figure 2.8). This bio-conjugation has provided a nidus for soft tissue integration with the oral epithelial and gingival cells. Thus, significantly reducing the soft tissue-related inflammation of the peri-implant tissue.

FTIR also aids in diagnostic prediction of squamous cell carcinoma with the aid of patient's saliva. When oral cancer develops, it leads to

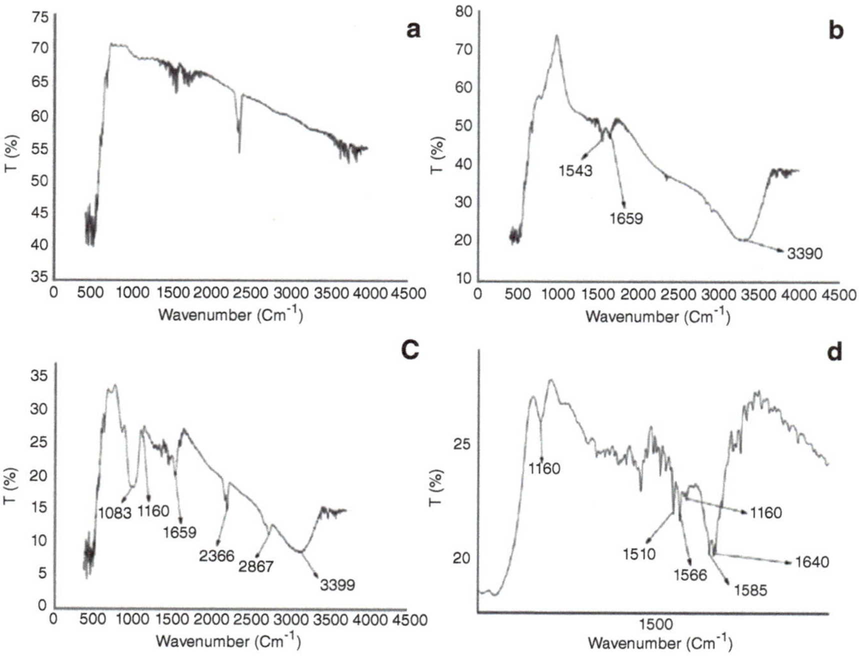

FIGURE 2.8 FTIR spectroscopy plots of grade V titanium alloy surfaces. (a) Polished Ti6Al4V. (b) Formation of hydroxyl group on alkali-treated surface. (c) Attachment of silane coupling agent (APTES). (d) Bio-conjugation of saline-treated surface with type I human collagen.

noticeable alterations in the secondary structure of proteins. The accuracy of FTIR to classify cancer tissues from normal subjects was 91.66% with 83.33% sensitivity and 100% specificity and precision.[49]

These studies highlight the usage of FTIR in dentistry as a reliable, non-invasive, reagent-free, and highly sensitive method to screen and analysis various dental materials.

2.4 ADVANTAGES AND LIMITATIONS OF FTIR

2.4.1 Advantages

FTIR offers certain advantages over the other characterization tools, such as the following: due to its high sensitivity, FTIR spectroscopy is able to detect trace amounts of the chemical species in the given sample (qualitative and quantitative assessment). It can be used to analyse a broad range of samples which are present in various phases such as solid, liquid, and gas. This technique is non-destructive in nature, i.e., the sample under the study is not altered or destroyed during the analysis. Making this a tool of choice for chemical analysis of fragile or valuable samples. Due to its very nature, technique results about the molecular characterization are obtained almost immediately.

2.4.2 Limitations

Like most of the tools and techniques, FTIR also possesses a few limitations, which might undermine its importance and applicability. It is possible to identify the functional groups in a given sample, but individual molecules and complex chemical structure and nature of compound cannot be determined. Due to this, it can be challenging to identify the complex samples, which contain mixture or overlapping spectra. Ideally, the sample under investigation should be dried properly, as the presence of water in the sample can interfere with the FTIR analysis. FTIR assessment of the samples requires the preparation of the thin film; this can be time-consuming and many a time requires technical expertise from the person who is handling the system. Furthermore, impurities/contaminations and interference from other molecules might lead to spectral interference; thus, the outcome might be inaccurate.

2.5 CONCLUSION

The FTIR technique is widely used for qualitative assessment of various functional groups in various fields such as food science, agriculture, chemical engineering, and dental materials. Dental materials are widely used as replacements for decayed, fractured, or misplaced dental hard and soft tissues. FTIR acts as a tool for estimations of various chemical natures, functionalities, and kinetics of the dental materials during its developmental phases as well as to estimate its clinical performances for any further modifications to make it more clinically useful.

REFERENCES

1. Blatz MB, Chiche G, Bahat O, *et al.* Evolution of aesthetic dentistry. *Journal of Dental Research* 2019;98:1294–1304.
2. Saeed F, Muhammad N, Khan AS, *et al.* Prosthodontics dental materials: From conventional to unconventional. *Materials Science & Engineering. C, Materials for Biological Applications* 2020;106:110167.
3. Ferracane JL. *Materials in dentistry: principles and applications*. Lippincott Williams & Wilkins: Philadelphia, USA; 2001.
4. Anusavice KJ, Shen C, Rawls HR, editors. *Phillips' science of dental materials*. Elsevier Health Sciences: St. Louis, USA; 2021.

5. Bonsor SJ, Pearson G. *A clinical guide to applied dental materials.* Elsevier Health Sciences: UK; 2012.
6. O'Brien WJ, editor. Dental materials and their selection. Chicago: Quintessence; 2002.
7. Barazanchi A, Li KC, Al-Amleh B, *et al.* Additive technology: Update on current materials and applications in dentistry. *Journal of Prosthodontics* 2017;26:156–163.
8. Van Noort R, Barbour ME. *Introduction to Dental Materials-E-Book: Introduction to Dental Materials-E-Book.* Elsevier Health Sciences: UK; 2023.
9. Sharan J, Singh S, Lale SV, *et al.* Applications of nanomaterials in dental science: A review. *Journal of Nanoscience and Nanotechnology* 2017;17: 2235–2255.
10. Dos Santos V, Brandalise RN, Savaris M. *Engineering of biomaterials.* New York, USA: Springer; 2017.
11. Marshall GW Jr, Marshall SJ, Bayne SC. Restorative dental materials: Scanning electron microscopy and x-ray microanalysis. *Scanning Microscopy* 1988;2:2007–2028.
12. Dusevich V, Melander JR, Eick JD. 13 SEM in dental research. *Scanning Electron Microscopy for the Life Sciences.* 2012; 6:211.
13. Burgo TL, Pereira GKR, Iglesias BA, *et al.* AFM advanced modes for dental and biomedical applications. *Journal of the Mechanical Behavior of Biomedical Materials* 2022;136:105475.
14. Kakaboura A, Fragouli M, Rahiotis C, *et al.* Evaluation of surface characteristics of dental composites using profilometry, scanning electron, atomic force microscopy and gloss-meter. *Journal of Materials Science. Materials in Medicine* 2007;18:155–163.
15. Kaczmarek K, Leniart A, Lapinska B, *et al.* Selected spectroscopic techniques for surface analysis of dental materials: A narrative review. *Materials (Basel)* 2021;14:2624.
16. Hany RA. Tools used for dental materials characterization. *Biomaterials Journal* 2022;1:9–16.
17. Kang BS, Sul YT, Oh SJ, *et al.* XPS, AES and SEM analysis of recent dental implants. *Acta Biomaterialia* 2009;5:2222–2229.
18. Ameen AP, Short RD, Johns R,*et al.* The surface analysis of implant materials. 1. The surface composition of a titanium dental implant material. *Clinical Oral Implants Research* 1993;4(3):144–150.
19. Sharan J, Koul V, Dinda AK,*et al.* Tethering of medical grade V titanium alloy surface through self-assembled monolayers (SAMS) for biomedical applications. *Applied Surface Science* 2017:412: 648–656.
20. Sharan J, Koul V, Dinda AK, *et al.* Bio-functionalization of grade V titanium alloy with type I human collagen for enhancing and promoting human periodontal fibroblast cell adhesion – an in-vitro study. *Colloids and Surfaces. B, Biointerfaces* 2018;161:1–9.
21. Griffiths PR, de Hasedh JA. *Fourier Transform Infrared Spectroscopy.* Wiley: USA;2006.
22. Smith BC. *Fundamental of Fourier Transform Infrared Spectroscopy.* USA, CRC Press; 2011.

23. Bunaciu AA, Aboul-Enein HY, Fleschin S. Application of Fourier transform infrared spectrophotometry in pharmaceutical drugs analysis. *Applied Spectroscopy Reviews* 2010;45:206–219.
24. Wessel E, Vogel C, Kolomiets O, *et al. FT-IR and NIR Spectroscopic Imaging: Principles, Practical Aspects and Applications in Material and Pharmaceutical Sciences. Infrared and Raman spectroscopic imaging*, 2nd edition. John Willey and Sons Inc.: Germany. 2009.
25. Smith BC. *Fundamental of Fourier Transform Infrared Spectroscopy*, 2nd edition. CRC Press: Boca Raton, USA. 2011.
26. Griffiths PR, de Haseth JA, Winefordner JD. *Fourier Transform Infrared Spectrometry*, 2nd edition. John Willey and Sons Inc.: USA. 2007.
27. Moore E (Edi). *Fourier Transform Infrared Spectroscopy (FTIR): Methods, analysis and research insight (Chemical engineering methods and technology)*. Nova Science Pub Inc. UK. 2016.
28. Seredin PV, Al-Zubaidi AA, Ippolitov YA. 2013 Research hydroxyapatite crystals and organic components of hard tooth tissues affected by dental caries using FTIR-microspectroscopy and XRD microdiffraction. *(Kondensirovannye sredy i mezhfaznye granitsy)2013*;15:224–231.
29. Orsini G, Orilisi G, Notarstefano V, *et al.* Vibrational imaging techniques for the characterization of hard dental tissues: From bench-top to chair-side. *Applied Sciences* 2021; 11:11953.
30. Almhöjd US, Norén JG, Arvidsson A, *et al.* Analysis of carious dentine using FTIR and ToF-SIMS. *Oral Health and Dental Management* 2014;13:735–744.
31. Liu Y, Yao X, Liu YW, *et al.* A Fourier transform infrared spectroscopy analysis of carious dentin from transparent zone to normal zone. *Caries Research* 2014;48:320–329
32. Puppin-Rontani J, Fugolin APP, Costa AR, *et al.* In vitro performance of 2-step, total etch adhesives modified by thiourethane additives. *International Journal of Adhesion and Adhesives* 2020;103:102688.
33. Shim JS, Lee SY, Song S-Y, *et al.* Polymerization efficiency of dental dual-cured resin cement light-cured at various times after the initiation of chemical activation. *International Journal of Polymeric Materials and Polymeric Biomaterials* 2020;69:622–628.
34. Fugolin AP, Lewis S, Logan MG, *et al.* Methacrylamide–methacrylate hybrid monomers for dental applications. *Dental Materials* 2020;36:1028–1037.
35. Fugolin A.P., Dobson A., Ferracane J.L., *et al.* Effect of residual solvent on performance of acrylamide-containing dental materials. *Dental Materials* 2019;35:1378–1387.
36. Fugolin APP, Navarro O, Logan MG, *et al.* Synthesis of di- and triacrylamides with tertiary amine cores and their evaluation as monomers in dental adhesive interfaces. *Acta Biomaterialia* 2020;115:148–159.
37. Fugolin AP, Dobson A, Huynh V, *et al.* Antibacterial, ester-free monomers: Polymerization kinetics, mechanical properties, biocompatibility and antibiofilm activity. *Acta Biomaterialia* 2019;100:132–141.
38. Delgado AH, Young AM. Modelling ATR-FTIR spectra of dental bonding systems to investigate composition and polymerisation kinetics. *Materials* 2021; 14:760.

39. Aftab M, Javed F, Haider S, *et al.* Design and Characterization of Chitosan-Based Smart Injectable Hydrogel for Improved Sustained Release of Antinarcotics. *Pharmaceuticals (Basel)* 2024;17:749.
40. ElBakry HA, Ammar MM, Moussa TA. Effect of nanodiamonds surface deposition on hydrophilicity, bulk degradation and*in-vitro*cell adhesion of 3D-printed polycaprolactone scaffolds for bone tissue engineering. *Biomedical Materials* 2024;19:10. 1088/1748-605X/ad5bac
41. Dos Santos Menezes L, Navarro da Rocha D. *et al.* Cellulose acetate scaffold coated with a hydroxyapatite/graphene oxide nanocomposite for application in tissue engineering. *Proceedings of the Institution of Mechanical Engineers. Part H* 2024;238:793–802.
42. Aminmansour S, Gomes de Carvalho AB, Medeiros Cardoso L, *et al.* Strontium-Doped Bioglass-Laden Gelatin Methacryloyl Hydrogels for Vital Pulp Therapy. *Journal of Functional Biomaterials* 2024;15:105.
43. Balto H, Bekhit MS, Auda SH, *et al.* Synergistic effect of Salvadora persica and chitosan nanoparticles against oropharyngeal microorganisms. *Scientific Reports* 2024;14:12997.
44. Daood U, Ilyas MS, Ashraf M, *et al.* A Novel Coated Suture Displays Antimicrobial Activity Without Compromising Structural Properties. *Journal of Oral and Maxillofacial Surgery* 2024;82:1147–1162.
45. Ali TW, Gul H, Fareed MA, *et al.* Biochemical properties of novel Carbon nanodot-stabilized silver nanoparticles enriched calcium hydroxide endodontic sealer. *PLoS One* 2024;3;19:e0303808.
46. Leenutaphong N, Phantumvanit P, Young AM, *et al.* Evaluation of setting kinetics, mechanical strength, ion release, and cytotoxicity of high-strength glass ionomer cement contained elastomeric micelles. *BMC Oral Health* 2024 20;24:713.
47. Zahid H, Shamim S, Kawish M, *et al.* Synthesis and characterization of honey based hybrid dental implants with enhanced antibacterial activity. *Pakistan Journal of Pharmaceutical Sciences* 2024;37:429–434.
48. Sharan J, Koul V, Dinda AK, *et al.* Bio-functionalization of grade V titanium alloy with type I human collagen for enhancing and promoting human periodontal fibroblast cell adhesion—An in-vitro study. *Colloids and Surfaces* 2018;161:1–9.
49. Shree P, Aggarwal Y, Kumar M, *et al.* Saliva based diagnostic prediction of oral squamous cell carcinoma using FTIR spectroscopy. *Indian Journal of Otolaryngology and Head & Neck Surgery* 2024;76:2282–2289.

CHAPTER 3

Infrared Spectroscopy and Fourier Transform Infrared Spectroscopy for Analysis of Dental Materials

Umut Pamukcu

3.1 INTRODUCTION

Spectroscopy techniques are classified according to the wavelength range of the electromagnetic radiation used. The region in the electromagnetic spectrum where light of different wavelengths or energy causes vibration in molecules as a result of its interaction with matter is called the infrared (IR) region. IR spectroscopy lies between the visible and microwave regions in the electromagnetic spectrum. IR spectroscopy is an analytical technique used to investigate the chemical composition of materials and their effects on biological structures such as body components (Stuart 2004). Since the characteristic vibrations of the components of the materials can be investigated with the IR light used in this technique, chemical specificity is naturally provided (Beddoe et al. 2023).

IR spectroscopy is a common technique used to characterize chemical composition of materials (both organic and inorganic) and evaluate their purity. Consequently, IR spectroscopy enables the identification of chemical components of multi-material systems. In IR spectroscopy, since only

DOI: 10.1201/9781003506805-3

low-energy photons interact with the sample and all frequencies are measured simultaneously, IR spectra can be obtained quickly, non-invasively, and without causing destruction (Beddoe et al. 2023). In addition, its advantages include having a cheap cost, sample preparation being done in a short time, the interpretation of the results not taking time, and the devices being easy to use (Preoteasa et al. 2018; Dutta 2017). Besides all these, the ability to test small amounts of the material and the ability of the light beam to penetrate a sample depth of approximately 0.5–3 μm are also important (Munajad, Subroto, and Suwarno 2018). These advantages have paved the way for the use of IR spectroscopy in various fields, such as polymer science, biochemistry, energy, and environmental toxicology (Wojda 2015; Duhn et al. 2021; Deb and Chana 2015; Orsini, Pagella, and Mitsiadis 2018). IR spectroscopy is also used to research materials in dentistry (Guglielmotti, Olmedo, and Cabrini 2019; Zhang and Kelly 2017; Della Bona et al. 2021).

Many IR spectrophotometer devices are produced for different application areas, depending on the spectrum regions and optical system differences. An IR interferogram's mathematical Fourier transform (FT) gives the IR spectrum. The process is known as FTIR. FTIR spectroscopy is commonly used to investigate gas, liquid, or solid-phase materials. Its basic working logic is to examine the interaction between electromagnetic radiation and the natural vibrations of the chemical bonds between the atoms of the material to be examined. Thus, it allows the determination of the characteristic vibrations of the atomic groups of the examined compound (Sacher and França 2018). In FTIR, the absorption of IR radiation paves the way for the transition from a low-energy state to a higher-energy state or an excited state (Parikh and Chorover 2005). The characteristic nature of the FTIR spectrum for a particular substance paves the way for IR spectrophotometric analysis to be generally used for qualitative analysis. Analysis with FTIR spectroscopy can characterize a pure substance, identify substances in compound form, and examine the interactions between atomic groups of materials formed as mixtures of different compounds. FTIR spectroscopy allows for determining characteristic vibrations of groups present in a compound (Lopes et al. 2018).

The use of IR and FTIR spectroscopies in dental materials, among the many spectroscopy techniques used today to analyse many materials, including dental tissues and synthetic biomimetic dental materials, will be the subject of this chapter.

3.2 AREAS OF USE OF INFRARED AND FOURIER TRANSFORM INFRARED SPECTROSCOPY IN DENTISTRY

Located between visible light and microwaves in the spectrum, IR radiation is divided into three regions: near-IR 12800–4000 cm^{-1}, mid-IR 4000–200 cm^{-1}, and far-IR 200–100 cm^{-1} (Lopes et al. 2018). In dentistry research, samples in the middle of the IR region (approximately 4000 to 400 cm^{-1}) have generally been evaluated (Chen et al. 2015). In dentistry, IR spectroscopy has typically been used to detect fluorosis and dental caries and to investigate the polymerization kinetics of adhesive and restorative materials and their effects on bacterial growth (Kaczmarek et al. 2021). In addition, with IR spectroscopy, the role of intermolecular collagen cross-linking, which is essential in the mechanical strength of dentin, and the structure of the materials used in pulp regeneration are evaluated. It is also a technique used to analyse biomimetic materials that mimic the organic or inorganic structures of the tooth, such as dentin and enamel tissues (Alania et al. 2020; Seredin et al. 2020).

3.2.1 Examination of Mineralized Dental Tissues with IR Spectroscopy

The mineralized tissues that comprise the tooth, such as enamel, dentin, and surrounding bone tissues, consist of water, organic, and mineral phases. The mineral matrix of tooth and related hard tissues consists mainly of calcium and phosphate, which form hydroxyapatite (HA) crystals, and trace amounts of potassium, bicarbonate, magnesium, sodium, and citrate. Differences in chemical composition can be observed depending on depth within the same tissue or between individuals (Bachmann et al. 2003; Khanarian et al. 2014; Donnelly et al. 2012). These hard tissues' chemical components, such as phosphate, water, collagen, and carbonate, strongly absorb the IR region. The values remain pretty similar, although the exact identification of the chemical components depends on some details, such as sample preparation or the age of the organism from which the sample was taken (Khanarian et al. 2014).

3.2.1.1 Examination of Enamel with IR Spectroscopy

Enamel, which covers the crown surface of the teeth and is the physically hardest tissue in the body, consists of 96% inorganic mineralized structures, 1% organic structures, and 3% water by weight (Duverger, Beniash, and Morasso 2016). HA crystals constitute a large part of the inorganic structure. The remaining parts consist of proteins, lipids, and water. Enamel tissue histologically shows a prismatic structure. Prisms are cylindrical and extend from the enamel-dentin junction to the tooth surface. Proteins, carbohydrates, lipids, and water are in the interprismatic region (Duverger, Beniash, and Morasso 2016).

Pathologies such as hypomaturation, erosion, and hypomineralization that can be observed in enamel were examined by FTIR spectroscopy (Kim et al. 2016; Sa et al. 2014; Sari et al. 2014; Taube, Marczewski, and Norén 2015). Enamel erosion caused by acids taken with acidic foods or beverages leads to enamel demineralization (Grippo, Simring, and Schreiner 2004). FTIR spectroscopy is highly sensitive in detecting changes in enamel erosion where phosphate content changes (Kim et al. 2016; Sari et al. 2014). Amelogenesis imperfecta (hypomaturation) is an inherited enamel disorder with a wide range of clinical symptoms, including a pale, opaque, discoloured appearance of the enamel, dentin-like radiodensity, and affected teeth that are highly susceptible to caries (Gadhia et al. 2012). FTIR spectroscopy results of hypomature enamel were reported as a higher matrix/mineral ratio, lower crystallization, and higher relative carbonate content (Sa et al. 2014). Molar Incisor Hypomineralization (MIH) is a developmental disorder in which the enamel on permanent first molars and incisors is affected. Hypomineralization zones in enamel are characterized by decreased mineral phase and larger amide bands in FTIR spectroscopy examinations (Taube, Marczewski, and Norén 2015).

Substances used to prevent enamel demineralization and remineralize the enamel were investigated by FTIR spectroscopy (Chung and Li 2013; Mohammed et al. 2014). Studies have shown that these substances (asparagine-serine-serine peptide and zinc) effectively reduce demineralization or increase remineralization on enamel surfaces, mainly in the phosphate regions in the HA structure. Lasers that increase enamel acid resistance to prevent dental caries are another topic investigated by FTIR spectroscopy (Lopes et al. 2018). CO_2, Nad:YAG and Er, and Cr:YSGG lasers have been proven effective in controlling dental caries (Lopes et al. 2018).

Another of the most researched topics using FTIR spectroscopy is teeth bleaching. Studies report that protein concentrations and carbonate and phosphate contents decrease proportionately to the treatment time and peroxide concentration (Lopes et al. 2018).

3.2.1.2 Examination of Dentin with IR Spectroscopy

Dentin, which constitutes the majority of dental tissue, is an elastic, hard, avascular, mineralized tissue that supports the enamel and surrounds the central pulp chamber. By weight, it contains approximately 70% mineral, 20% organic matrix, and 10% liquid (Deng et al. 2013). The main component of the dentin organic matrix is fibrillar type I collagen (90%), which serves as a scaffold for mineral crystal deposition and deposition of non-collagenous matrix. The accumulation of minerals within and between collagen fibrils forms a rigid, flexible, and resistant matrix. In histological dentin architecture, peritubular dentin, dentinal tubules, and intertubular dentin are observed. Dentin composition can be altered by physiological and pathological processes that cause changes in its biological behaviour and structure (Marshall et al. 2001).

Correlating the molecular structure of dentin with its mechanical, microscopic, and histological aspects is essential for a complete understanding of its properties and role (Liu et al. 2014). In this respect, FTIR spectroscopy is a powerful method that produces information directly about the chemical composition of dentin samples and indirectly about its molecular structure. Studies have used FTIR spectroscopy to investigate the effect of lasers and plasma on dentin (Lopes et al. 2018). Collagen loss after laser and plasma treatment has increased adhesive/dentin interface bonding (Lopes et al. 2018).

The effect of irrigation on the chemical interaction between root canal sealer and dentin and the characterization of caries formation in dentin were investigated through FTIR spectroscopy (Liu et al. 2014; Maske et al. 2015; Neelakantan et al. 2015).

With ATR-FTIR spectroscopy, an IR spectroscopy method, the role of intermolecular collagen cross-linking in the mechanical behaviour of the dentin extracellular matrix and the structural and biochemical changes of dentin collagen before and after protein destabilization were investigated (Alania et al. 2020).

3.2.2 Examination of Synthetic Dental Materials with IR Spectroscopy

IR spectroscopy monitors their polymerization kinetics to improve the mechanical properties of resin-based composites and adhesives used in dentistry (Kaczmarek et al. 2021). In particular, the widely used FTIR spectroscopy determines the structural characterization of many dental materials, including implant and coating materials, biopolymers, bioceramics, resin-based composites, adhesives, bioglass, and some self-hardening dental materials (Kaczmarek et al. 2021). Similarly, ATR-FTIR and diffuse reflectance IR Fourier transform spectroscopy (DRIFTS) techniques, which can be considered variations of the IR spectroscopy, were used in the analysis of dental materials such as resin-based composites, adhesives, stainless steel, titanium implants, and HA used in their coating (Lach et al. 2020; Jonauske et al. 2019; Rigo et al. 2004; Sacher and França 2018).

Some of the most important problems encountered in using new dental biomaterials are related to surface phenomena that occur at their interface with oral tissues and other environmental structures. An issue at the interface may cause material leakage, where most harmful agents penetrate the tooth or the structure of the dental biomaterial, damaging both components and limiting their lifespan. To prevent this phenomenon, examining the materials used in dentistry with IR spectroscopy provides essential data.

Photopolymerization, used in dentistry to harden various dental materials such as resin-based composites, is affected by the components of the material. Among the factors that limit the clinical lifespan of composite restorations, the susceptibility of methacrylates to hydrolytic and enzymatic degradation is an essential factor. As a result of the evaluation of polymerization kinetics with IR spectroscopy, hybrid versions show a lower degree of conversion (DC) and polymerization rate. In contrast, methacrylate controls showed the highest values (Fugolin et al. 2020). With IR spectroscopy measurements, significant differences were observed in the DC when the groups cured quickly with light, and those with a longer delay in light curing were compared (Shim et al. 2020). Studies using molecular multidimensional IR spectroscopy have also been conducted to examine the integration between the new generation biomimetic composite resin, which replicates the mineral-organic complexes of the tooth (enamel and dentin). Microspectroscopy data demonstrate the existence of organic and mineral interactions at the interface between biomimetic materials and natural hard tissues of the human tooth (Seredin et al. 2020).

In dental adhesive resin systems, the presence during polymerization of solvents such as ethanol, water, and acetone, which reduce the resin viscosity and enable relatively viscous and hydrophobic monomers to penetrate the hydrophilic dentin substrate, dramatically affects the reaction kinetics. Once the adhesive material reaches the intended depth, air drying must remove the solvent entirely (Sauro et al. 2007). However, in clinical practice, 12–14% of the solvent by weight remains in the environment after air drying (Garcia et al. 2010; Yiu et al. 2005). When the DC was monitored in IR spectroscopy, the presence of residual solvent affected the polymerization kinetics and mechanical properties of the monomer system (Fugolin et al. 2019).

Calcium phosphate (CaP) bioceramics are increasingly used due to their chemical similarity to the inorganic component of mammalian bone and teeth. They are frequently used in medicine and dentistry because they are non-toxic (Mobasherpour, Heshajin, and Zakeri 2007; Sato et al. 2006). Its uses include orthopaedic applications such as coating implants to increase biocompatibility, filling bone defects, and creating a bone tissue skeleton (Abidi and Murtaza 2014; Sato et al. 2006; Dias et al. 2008). In an FTIR spectroscopy study for phase identification, CaP phase precipitation was highly pH-dependent and not dependent on the initial Ca/P ratio (Rafeek, Choi, and Evans 2018).

Polyetheretherketone (PEEK), which has a bone-like hardness, excellent fatigue resistance, and minimal immune toxicity, appears to have properties that could be used as an orthopaedic and dental implant material. The surface modification exhibited by PEEK becomes an important option, especially when excellent osseointegration is desired. A study conducted with ATR-FTIR spectroscopy to investigate the surface characterization of the PEEK sample reported that this method has great potential (Yakufu et al. 2020).

Bioactive Glass Ceramics (BGC), which resemble bone components, have recently become one of the most promising biomaterials for treating bone defects, including dentistry. Therefore, it has been widely applied as a filling material for bone tissue regeneration. In the study, where the intermolecular interactions of the particles in the sample were determined by FTIR spectroscopy analysis, it was shown that molecular bonds were formed between the other materials used and the ceramics (Kumar et al. 2018).

REFERENCES

Abidi, Syed Sibte Asghar, and Qasim Murtaza. 2014. "Synthesis and characterization of nano-hydroxyapatite powder using wet chemical precipitation reaction." *Journal of Materials Science & Technology* 30 (4): 307–310.

Alania, Yvette, Mariana Cavalcante dos Reis, Joo-Won Nam, Rasika S Phansalkar, James McAlpine, Shao-Nong Chen, Guido F Pauli, and Ana K Bedran-Russo. 2020. "A dynamic mechanical method to assess bulk viscoelastic behavior of the dentin extracellular matrix." *Dental Materials: Official Publication of the Academy of Dental Materials* 36 (12): 1536.

Bachmann, Luciano, Rolf Diebolder, Raimund Hibst, and Denise Maria Zezell. 2003. "Infrared absorption bands of enamel and dentin tissues from human and bovine teeth." *Applied Spectroscopy Reviews* 38 (1): 1–14.

Beddoe, Max, Thorsten Gölz, Martin Barkey, Enrico Bau, Matthias Godejohann, Stefan A Maier, Fritz Keilmann, Marioara Moldovan, Doina Prodan, and Nicoleta Ilie. 2023. "Probing the micro-and nanoscopic properties of dental materials using infrared spectroscopy: A proof-of-principle study." *Acta Biomaterialia* 168: 309–322.

Chen, Yanyan, Caineng Zou, Maria Mastalerz, Suyun Hu, Carley Gasaway, and Xiaowan Tao. 2015. "Applications of Micro-Fourier Transform Infrared Spectroscopy (FTIR) in the Geological Sciences--A Review." *International Journal of Molecular Sciences* 16 (12): 30223–30250.

Chung, Hsiu-Ying, and Cheng Che Li. 2013. "Microstructure and nanomechanical properties of enamel remineralized with asparagine-serine-serine peptide." *Materials Science & Engineering. C, Materials for Biological Applications* 33 (2): 969–973.

Deb, Sanjukta, and Simran Chana. 2015. "Biomaterials in relation to dentistry." *Frontiers of Oral Biology* 17: 1–12.

Della Bona, Alvaro, Viviane Cantelli, Vitor T. Britto, Kaue F. Collares, and Jeffrey W. Stansbury. 2021. "3D printing restorative materials using a stereolithographic technique: a systematic review." *Dental Materials: Official Publication of the Academy of Dental Materials* 37 (2): 336–350.

Deng, Meng, Xiaoli Dong, Xuedong Zhou, Lufei Wang, Hong Li, and Xin Xu. 2013. "Characterization of dentin matrix biomodified by Galla chinensis extract." *Journal of Endodontics* 39 (4): 542–547.

Dias, Susana, Vera Lourenço, Carlos A Nogueira, and Fernando A Costa Oliveira. 2008. "Contrasting morphology of calcium phosphate powders prepared by the wet method using poly (ethylene glycol)." *Materials Science Forum* 587: 32–36.

Donnelly, Eve, Dennis S Meredith, Joseph T Nguyen, and Adele L Boskey. 2012. "Bone tissue composition varies across anatomic sites in the proximal femur and the iliac crest." *Journal of Orthopaedic Research* 30 (5): 700.

Duhn, Christopher, Ghadeer Thalji, Sandra Al-Tarwaneh, and Lyndon F Cooper. 2021. "A digital approach to robust and esthetic implant overdenture construction." *Journal of Esthetic and Restorative Dentistry* 33 (1): 118–126.

Dutta, Aastha. 2017. "Fourier Transform Infrared." *Spectroscopic Methods for Nanomaterials Characterization* 2: 73.

Duverger, Olivier, Elia Beniash, and Maria I Morasso. 2016. "Keratins as components of the enamel organic matrix." *Matrix Biology: Journal of the International Society for Matrix Biology* 52: 260–265.

Fugolin, Ana P., Adam Dobson, Jack L. Ferracane, and Carmem S. Pfeifer. 2019. "Effect of residual solvent on performance of acrylamide-containing dental materials." *Dental Materials: Official publication of the Academy of Dental Materials* 35 (10): 1378–1387.

Fugolin, Ana P., Steven Lewis, Matthew G. Logan, Jack L. Ferracane, and Carmem S. Pfeifer. 2020. "Methacrylamide-methacrylate hybrid monomers for dental applications." *Dental Materials: Official Publication of the Academy of Dental Materials* 36 (8): 1028–1037.

Gadhia, K., S. McDonald, N. Arkutu, and K. Malik. 2012. "Amelogenesis imperfecta: an introduction." *British Dental Journal* 212 (8): 377–379.

Garcia, Georges, Karen Barros Parron Fernandes, Fernanda Cristina Pimentel Garcia, Paulo Henrique Perlatti D'Alpino, Nádia da Rocha Svizero, and Linda Wang. 2010. "Solvent retention of contemporary commercial dentin bonding agents in a demineralized dentin matrix." *European Journal of Dentistry* 4 (3): 293–297.

Grippo, John O., Marvin Simring, and Steven Schreiner. 2004. "Attrition, abrasion, corrosion and abfraction revisited: a new perspective on tooth surface lesions." *Journal of the American Dental Association (1939)* 135 (8): 1109–1165.

Guglielmotti, María B., Daniel G. Olmedo, and Rómulo L. Cabrini. 2019. "Research on implants and osseointegration." *Periodontology 2000* 79 (1): 178–189.

Jonauske, Vilma, Sandra Stanionyte, Shih-Wen Chen, Aleksej Zarkov, Remigijus Juskenas, Algirdas Selskis, Tadas Matijosius, Thomas C.K. Yang, Kunio Ishikawa, and Rimantas Ramanauskas. 2019. "Characterization of sol-gel derived calcium hydroxyapatite coatings fabricated on patterned rough stainless steel surface." *Coatings* 9 (5): 334.

Kaczmarek, Katarzyna, Andrzej Leniart, Barbara Lapinska, Slawomira Skrzypek, and Monika Lukomska-Szymanska. 2021. "Selected Spectroscopic Techniques for Surface Analysis of Dental Materials: A Narrative Review." *Materials (Basel, Switzerland)* 14 (10): 2624.

Khanarian, Nora T., Margaret K. Boushell, Jeffrey P. Spalazzi, Nancy Pleshko, Adele L. Boskey, and Helen H. Lu. 2014. "FTIR-I Compositional Mapping of the Cartilage-to-Bone Interface as a Function of Tissue Region and Age." *Journal of Bone and Mineral Research: The Official Journal of the American Society for Bone and Mineral Research* 29 (12): 2643.

Kim, In-Hye, Jun Sik Son, Bong Ki Min, Young Kyoung Kim, Kyo-Han Kim, and Tae-Yub Kwon. 2016. "A simple, sensitive and non-destructive technique for characterizing bovine dental enamel erosion: attenuated total reflection Fourier transform infrared spectroscopy." *International journal of oral science* 8 (1): 54–60.

Kumar, S. Dinesh, K. Mohamed Abudhahir, N. Selvamurugan, S. Vimalraj, R. Murugesan, N. Srinivasan, and A. Moorthi. 2018. "Formulation and biological actions of nano-bioglass ceramic particles doped with Calcarea phosphorica for bone tissue engineering." *Materials Science & Engineering. C, Materials for Biological Applications* 83: 202–209.

Lach, Sławomir, Przemysław Jurczak, Natalia Karska, Agnieszka Kubiś, Aneta Szymańska, and Sylwia Rodziewicz-Motowidło. 2020. "Spectroscopic Methods Used in Implant Material Studies." *Molecules (Basel, Switzerland)* 25 (3): 579.

Liu, Y., X. Yao, Y.W. Liu, and Y. Wang. 2014. "A Fourier transform infrared spectroscopy analysis of carious dentin from transparent zone to normal zone." *Caries Research* 48 (4): 320–329.

Lopes, Camila De Carvalho Almança, Pedro Henrique Justino Oliveira Limirio, Veridiana Resende Novais, and Paula Dechichi. 2018. "Fourier transform infrared spectroscopy (FTIR) application chemical characterization of enamel, dentin and bone." *Applied Spectroscopy Reviews* 53 (9): 747–769.

Marshall, G.W., S. Habelitz, R. Gallagher, M. Balooch, G. Balooch, and S.J. Marshall. 2001. "Nanomechanical properties of hydrated carious human dentin." *Journal of Dental Research* 80 (8): 1768–1771.

Maske, Tamires T., Cristina P. Isolan, Françoise H. van de Sande, Aline C. Peixoto, André L. Faria-E-Silva, Maximiliano S. Cenci, and Rafael R. Moraes. 2015. "A biofilm cariogenic challenge model for dentin demineralization and dentin bonding analysis." *Clinical Oral Investigations* 19 (5): 1047–1053.

Mobasherpour, I., M. Soulati Heshajin, and M. Zakeri. 2007. "Synthesis of nanocrystalline hydroxyapatite by using precipitation method." *Journal of Alloys and Compounds* 430 (1–2): 330–333.

Mohammed, N.R., M. Mneimne, R.G. Hill, M. Al-Jawad, R.J.M. Lynch, and P. Anderson. 2014. "Physical chemical effects of zinc on in vitro enamel demineralization." *Journal of Dentistry* 42 (9): 1096–1104.

Munajad, Abi, Cahyo Subroto, and Suwarno. 2018. "Fourier transform infrared (FTIR) spectroscopy analysis of transformer paper in mineral oil-paper composite insulation under accelerated thermal aging." *Energies* 11 (2): 364.

Neelakantan, Prasanna, Subash Sharma, Hagay Shemesh, and Paul R. Wesselink. 2015. "Influence of Irrigation Sequence on the Adhesion of Root Canal Sealers to Dentin: A Fourier Transform Infrared Spectroscopy and Push-out Bond Strength Analysis." *Journal of Endodontics* 41 (7): 1108–1111.

Orsini, Giovanna, Pierfrancesco Pagella, and Thimios A Mitsiadis. 2018. "Modern Trends in Dental Medicine: An Update for Internists." *The American Journal of Medicine* 131 (12): 1425–1430.

Parikh, Sanjai J., and Jon Chorover. 2005. "FTIR spectroscopic study of biogenic Mn-oxide formation by Pseudomonas putida GB-1." *Geomicrobiology Journal* 22 (5): 207–218.

Preoteasa, Eugen A., Elena S. Preoteasa, Ioana Suciu, and Ruxandra N Bartok. 2018. "Atomic and nuclear surface analysis methods for dental materials: A review." *AIMS Materials Science* 5 (4): 781–844.

Rafeek, A.D., G. Choi, and L.A. Evans. 2018. "Morphological, spectroscopic and crystallographic studies of calcium phosphate bioceramic powders." *Journal of the Australian Ceramic Society* 54: 161–168.

Rigo, E.C.S., A.O. Boschi, M. Yoshimoto, S. Allegrini Jr, B. Konig Jr, and M.J. Carbonari. 2004. "Evaluation in vitro and in vivo of biomimetic hydroxyapatite coated on titanium dental implants." *Materials Science and Engineering: C* 24 (5): 647–651.

Sa, Yue, Shanshan Liang, Xiao Ma, Steven Lu, Zhejun Wang, Tao Jiang, and Yining Wang. 2014. "Compositional, structural and mechanical comparisons of normal enamel and hypomaturation enamel." *Acta Biomaterialia* 10 (12): 5169–5177.

Sacher, Edward, and Rodrigo França. 2018. "Surface Analysis Techniques for Dental Materials." *Dental Biomaterials* 2: 1.

Sari, Mustafa Erhan, Aliye Gediz Erturk, Alp Erdin Koyuturk, and Yunus Bekdemir. 2014. "Evaluation of the effect of food and beverages on enamel and restorative materials by SEM and Fourier transform infrared spectroscopy." *Microscopy Research and Technique* 77 (1): 79–90.

Sato, Kimiyasu, Yuji Hotta, Takaaki Nagaoka, Masaki Yasuoka, and Koji Watari. 2006. "Agglomeration control of hydroxyapatite nano-crystals grown in phase-separated microenvironments." *Journal of Materials Science* 41: 5424–5428.

Sauro, Salvatore, David H. Pashley, Marco Montanari, Stefano Chersoni, Ricardo M. Carvalho, Manuel Toledano, Raquel Osorio, Franklin R. Tay, and Carlo Prati. 2007. "Effect of simulated pulpal pressure on dentin permeability and adhesion of self-etch adhesives." *Dental Materials: Official Publication of the Academy of Dental Materials* 23 (6): 705–713.

Seredin, P.V., O.A. Uspenskaya, D.L. Goloshchapov, I. Yu Ippolitov, Jitraporn Pimm Vongsvivut, and Yu A. Ippolitov. 2020. "Organic-Mineral Interaction between Biomimetic Materials and Hard Dental Tissues." *Sovremennye tekhnologii v meditsine* 12 (1): 43–50.

Shim, Ji Suk, Soo Young Lee, So-Yeon Song, Nayansi Jha, and Jae Jun Ryu. 2020. "Polymerization efficiency of dental dual-cured resin cement light-cured at various times after the initiation of chemical activation." *International Journal of Polymeric Materials and Polymeric Biomaterials* 69, 622–628.

Stuart, Barbara H. 2004. *Infrared spectroscopy: fundamentals and applications.* John Wiley & Sons.

Taube, F., M. Marczewski, and J.G. Norén. 2015. "Deviations of inorganic and organic carbon content in hypomineralised enamel." *Journal of Dentistry* 43 (2): 269–278.

Wojda, Sylwia M. 2015. "Comparative analysis of two methods of assessment wear of dental materials." *Acta Mechanica et Automatica* 9 (2): 105–109.

Yakufu, Maihemuti, Zongliang Wang, Yu Wang, Zixue Jiao, Min Guo, Jianguo Liu, and Peibiao Zhang. 2020. "Covalently functionalized poly (etheretherketone) implants with osteogenic growth peptide (OGP) to improve osteogenesis activity." *RSC Advances* 10 (17): 9777–9785.

Yiu, Cynthia K.Y., Edna L. Pashley, Noriko Hiraishi, Nigel M. King, Cecilia Goracci, Marco Ferrari, Ricardo M. Carvalho, David H. Pashley, and Franklin R. Tay. 2005. "Solvent and water retention in dental adhesive blends after evaporation." *Biomaterials* 26 (34): 6863–6872.

Zhang, Yu, and J. Robert Kelly. 2017. "Dental ceramics for restoration and metal veneering." *Dental clinics of North America* 61 (4): 797–819.

CHAPTER 4

Ultraviolet and Visible Spectroscopy for Analysis of Dental Materials

Nebiha Gozde Ispir and Zühre Akarslan

4.1 INTRODUCTION

Ultraviolet and visible spectroscopy (UV-VIS spectroscopy) has been used for the analysis of the structure of molecules for a long time. It is capable of making fast and cheap qualitative and quantitative analyses. It can be applied to a wide range of chemicals. Liquids, solids, and gases could be analyzed with this technique (Abdolkarimi-Mahabadi, Bayat, and Mohammadi 2021; Kaczmarek et al. 2021).

The basic principle of this technique is the measurement of the absorption or reflection of a light beam in the ultraviolet and visible light spectrum from the examined material (Abdolkarimi-Mahabadi, Bayat, and Mohammadi 2021). The measurements are made with a UV-VIS spectrophotometer. Modern spectrophotometers have user-friendly software interfaces. This makes the technique even simpler to use. As the procedure is non-destructive, one can repeat the analysis of the same sample. This is beneficial during quality assurance and quality control (Wang and Chu 2013).

Spectroscopy studies the interaction between electromagnetic rays and matter. The wavelength of UV radiation is between 10 and 400 nm. The range 10–200 nm is the far UV region. Since the absorption of N_2 and O_2

DOI: 10.1201/9781003506805-4

gases is in the range of 10–200 nm, this process should be done under vacuum. Thus, it is also called the vacuum region. The range 200–400 nm is called the near UV or UV region. 400–800 nm is called the visible region. Thus, UV-VIS devices generally operate between 200 and 800 nm. A light source emitting a wide range of wavelengths should be used in this technique (Abdolkarimi-Mahabadi, Bayat, and Mohammadi 2021). Mostly, a single xenon lamp is capable as a light source as it has high intensity and could be used for the UV and VIS electromagnetic spectrum. However, xenon lamps are expensive and less stable compared with tungsten and halogen lamps. Some instruments use two lamps. Mostly, a tungsten and halogen lamp is used for a visible light source and a deuterium lamp for ultraviolet light (Diffey 2002).

In spectroscopy, a line spectrum is generated when an atom of a particular element is irradiated with energy, resulting in the excitation of the outer shell electrons to a higher state. Later, when these electrons come back to their ground state, they emit their excitation energies corresponding to a particular frequency. This presents a characteristic of an element.

When the energy required for the excitation of the electron is absorbed by molecule, this absorption is recorded as a UV-VIS spectrum with the UV-VIS device. The spectra are drawn in various units (such as absorbance and transmittance) as absorption intensity versus wavelength. The electronic energy levels in the ground and excited states also include vibrational and rotational energy levels. Since vibration and rotational energy levels are also excited during electronic excitation, the electronic excitation expected as an absorption line expands and turns into an absorption band.

4.2 TYPES OF ELECTRONIC TRANSITIONS IN UV-VIS SPECTROSCOPY

When electromagnetic radiation is absorbed by the sample, an electron transition occurs from the σ and π orbitals to σ^* or π^*. Most of the absorptions are $\pi \rightarrow \pi^*$, $n \rightarrow \sigma^*$, and $n \rightarrow \pi^*$ transitions (σ (sigma) electrons, π (pi) electrons, and non-bonding *n* electrons).

$\sigma \rightarrow \sigma^*$ transitions: These transitions are observed in the far UV region because they require very high energy levels. σ electrons belonging to C-C and C-H bonds make such transitions.

$n \rightarrow \sigma^*$ transitions: These transitions are observed in saturated compounds containing heteroatoms.

$\sigma \rightarrow \pi^*$ and $\pi \rightarrow \sigma^*$ transitions: Absorptions are weak at this kind of transition.

$\pi \rightarrow \pi^*$ transitions: These transitions are severe transitions. They are observed at unsaturated compounds.

$n \rightarrow \pi^*$ transitions: These transitions are almost seen at the far-right side of the spectrum.

The temperature, solvent, and molecule structure of the examined material affect electronic transitions (Abdolkarimi-Mahabadi, Bayat, and Mohammadi 2021; Pentassuglia, Agostino, and Tommasi 2018; Kaczmarek et al. 2021).

In this technique, the Beer–Lambert law (Beer's law) is used for the quantitative assessment of sample concentration. This law assumes that the concentration of specific components in the sample can be found with a calibration equation. This equation helps one to predict the quantities of the material in interest. The Beer–Lambert law determines a simple linear relationship between sample composition and the spectrum. The basic principle of this law is that the absorbed energy by the examined sample under a specific beam of the spectrophotometer is directly and linearly related to the amount of the component in the sample (Duckworth 1998).

The Beer–Lambert formula is expressed as follows:

$$A = \text{Log}_{10}\,(I_0/I) = \mathcal{E}CL$$

A = Absorbance measured

I_0 = Light intensity of the incident light at a specific wavelength

I = Transmitted intensity

L = Path length

C = Concentration of the absorbing spices

$\mathcal{E}$ = Constant

The constant in the formula is expressed as molar absorptivity/extinction coefficient of spices and wavelength (Wang and Chu 2013).

The measurements should be made under the same standards to obtain successful results. The spectrophotometer is useful for such applications. The pH or single-element atomic absorption methods only give single-point measurements for each calibration and unknown sample. However, the spectrophotometer provides measurements with multiple data points. This has many advantages. More measurements are made on the sample to use in generating the calibration equation. This increases the accuracy of the measurement.

The usage of UV-VIS spectroscopy for dental materials is presented in the literature. Some materials used in dentistry have specific characteristic absorption bands in the ultraviolet and visible light regions of the electromagnetic spectrum. The amount of carbon nanostructures in the examined material is correlated with the intensity of surface plasmon resonance. Though more intense surface plasma resonance shows higher concentrations of carbon nanostructures (Abdolkarimi-Mahabadi, Bayat, and Mohammadi 2021; Kaczmarek et al. 2021).

4.3 USE OF ULTRAVIOLET AND VISIBLE SPECTROSCOPY IN ANALYSIS OF DENTAL MATERIALS

UV-VIS spectroscopy is used in various areas of dentistry. This technique provides explicit quantitative and qualitative evaluations of dental materials (Kaczmarek et al. 2021). The main areas of use of UV-Vis spectroscopy in dental materials are as follows:

1. Evaluation of the effectiveness of teeth whitening (bleaching) materials
2. Evaluation of the antibacterial efficacy and structure of composite fillings
3. Evaluation of alternative herbal extracts to prevent dental caries and periodontitis.
4. Assessment of the opacity of materials used in fixed dental prostheses

4.4 EVALUATION OF THE EFFECTIVENESS OF TEETH WHITENING (BLEACHING) MATERIALS

4.4.1 Teeth Color and Causes of Teeth Discoloration

Tooth color combines the optical properties of the layers that make up the tooth structure (Ten Bosch and Coops 1995; Oktay 2006). Tooth color is observed due to the interaction between light and teeth in the form of transmission, reflection, absorption, and scattering of light through tooth tissues (Ten Bosch and Coops 1995; Oktay 2006). Dental tissues are enamel, dentin, and pulp. While the dentin determines the color of teeth, the enamel on it also contributes slightly to the determination of teeth color (Ten Bosch and Coops 1995; Oktay 2006).

Color changes can be seen in the teeth. These discolorations on the teeth are divided according to their etiology internal and external discolorations (Carey 2014).

External discoloration occur due to the penetration of materials such as tea, coffee, cigarettes, chlorhexidine, iron salts, and alcohol into the enamel surface (Carey 2014).

Internal discolorations occur due to reasons such as genetics, the appearance of yellow dentin due to the wear of the enamel with age, the use of tetracycline-derived antibiotics taken during the development of the teeth, and the consumption of water resources with high fluoride content (Carey 2014).

These discolorations on the teeth cause aesthetic anxiety in individuals. In addition to dental health, the appearance and color of teeth are important for people in terms of aesthetics. Many material studies have been done to whiten teeth for the aesthetic appearance of teeth since the 1800s (Frysh et al. 1995). Truman has used a solution known as Labarraqque, produced from chloride and acetic acid, to whiten a devital tooth in 1864 (Truman 1864). In a study conducted by Harlan in 1884, it was stated that hydrogen peroxide could be used for whitening purposes (Harlan 1884). Today's materials used for dental whitening are hydrogen peroxide and carbamide peroxide prepared in different concentrations. In addition, ozone gas has recently taken its place among the materials used in dental whitening treatment (Tessier et al. 2010).

Dental whitening treatment is performed with different methods for the etiology of discoloration. While vital teeth are called in-office bleaching and at-home bleaching, the whitening process for devital teeth is called walking bleaching, internal power bleaching, and internal/external bleaching technique.

4.4.2 Office Bleaching

It is a whitening technique applied to vital teeth in a clinical environment to achieve rapid whitening to remove external discolorations due to diet, discolorations caused by mild fluorosis, discolorations due to mild and moderate drug use, and age-related discolorations. In this technique, 25–45% hydrogen peroxide is used as a whitening agent (Kury et al.

2022). Lasers, light-emitting diodes (LEDs), plasma arc lamps (PACs), and halogen lamps are used to activate hydrogen peroxide by increasing the rate of its decomposition into free radicals (Maran et al. 2018).

4.4.3 Home Bleaching

Home bleaching was first described and implemented in 1989 (Haywood and Heymann 1989). It is a technique that the patient applies at home in accordance with the procedures given by the dentist and is controlled by the dentist (TÜZEL and CAN 2022). In this technique, carbamide peroxide, prepared with a concentration varying from 10–30%, is used as a whitening agent in the production of companies (TÜZEL and CAN 2022). Increasing the concentration of the whitening agent ensures a faster completion of the bleaching process. However, this situation may increase tooth sensitivity. On the contrary, the use of low-concentration carbamide peroxide provides less tooth sensitivity, but the bleaching process takes more time (Tüzel and Can 2022). Considering these situations, the most applicable and safest method for the patient is decided in cooperation with the physician and the patient. Home bleaching method under the supervision of a dentist has advantages such as low cost, short time spent in the clinic, and being safe for the patient to apply (Tüzel and Can 2022). In addition, it has disadvantages compared to office bleaching, such as the patient finding the application method difficult, having difficulty in adaptation, and achieving results in a longer period of time (Tüzel and Can 2022).

There are also home bleaching techniques that do not need to be applied under the supervision of a dentist. These can be applied by patients with the help of toothpaste and a toothbrush. For this method, low concentrations of hydrogen peroxide or carbamide peroxide are added to toothpaste and pastes as a whitening agent.

4.4.4 Bleaching for the Nonvital Teeth

As a result of teeth losing their vitality (nonvital tooth), tooth color may darken. Root canal treatment is applied for non-vital tooth treatment. During this treatment, discoloration of the teeth may occur. Blood products cause pink discoloration on the teeth due to bleeding that occurs during pulp extirpation in root canal treatment (Arens 1989; Attin, Paque, Ajam, and Lennon 2003). Then, the degradation products of blood, such as hemosiderin, hematin, and hemin, release iron during hemolysis. This released iron interacts with hydrogen sulfide produced by bacteria and turns into black ferric sulfide, which may cause the tooth to appear gray (Attin, Paque, Ajam, and Lennon 2003). Apart from this, the root canal filling materials may also cause color changes in the teeth.

Walking bleaching is the most commonly used method for whitening nonvital teeth. In this method, sodium perborate and distilled water are used as whitening agents. Whitening agents are placed in the pulp chamber of the tooth and covered with a temporary filling material at least 3 mm thick. The patient is given an appointment a few days later, and the effect of the whitening process is evaluated. The patient may repeat this procedure several times for adequate whitening (Alqahtani 2014; Usta and Gözükara 2023). Compared to alternative prosthetic treatments, this whitening method has the advantages of being minimally invasive, low cost, less time spent in the clinic, and being safe and comfortable for the patient (Spasser 1961; Zimmerli, Jeger, and Lussi 2010). However, sodium perborate has been banned in the European Union due to its carcinogenic and mutagenic effects since 2015 (Reitzer, Ehlinger, and Minoux 2019). For the walking bleaching method, it is recommended to use 10–16% carbamide peroxide in accordance with the European Union guidelines (Greenwall-Cohen and Greenwall 2019).

The internal power bleaching technique is a technique developed to increase the speed and effectiveness of whitening treatment. In this technique, 30–35% hydrogen peroxide gel is used as a whitening agent. A whitening agent is placed in the pulp chamber of the nonvital tooth, and the process is carried out by providing photooxidation or thermocatalytic activation in this technique (Başar and Çıkman, 2023; Plotino et al. 2008).

The other technique, used for whitening of nonvital teeth, is internal and external bleaching. In this technique, 10% carbamide peroxide is used as a whitening agent. In this method, a protective barrier is placed over the root canal filling to isolate the pulp chamber. A patient-specific whitening plate is prepared. With the help of the prepared plate, the whitening agent is applied to both the inner and outer surfaces of the tooth. The application of whitening agent is repeated by the patient. The entrance cavity opened to place the whitening agent is not closed in this method; it can only be closed with pellet cotton. This is a disadvantage of this method due to reasons such as food residues entering and cleaning not being ensured. In order to avoid disadvantages, patients are asked to pay attention to the cleaning of this area, and it is recommended that they cover it with a cotton pellet after each whitening procedure (Reitzer, Ehlinger, and Minoux 2019).

4.4.5 Mechanism of the Whitening Process

The whitening process is based on the chemical oxidation–reduction reaction between the coloring agent (reducing agent) and the whitening agent (oxidizing agent), which is formed by high molecular weight, complex organic molecules that reflect a certain wavelength of light (Punia Sandhya Kapoor 2015; Dahl and Pallesen 2003). As a result of the oxidative reaction, molecular oxygen is released, which can break down the pigments responsible for coloration, which penetrate into mineralized tooth tissues. Oxygen molecules enter the organic enamel-dentin matrix and decompose dark pigments by changing long molecular chains. Lower-weight molecules that reflect less light are formed. In this way, the coloration is reduced or completely eliminated (Punia Sandhya Kapoor 2015; Rotstein and Li 2008).

Whitening agents used today for this purpose have active and inactive ingredients (Alqahtani 2014).

Active ingredients are as follows:

- carbamide peroxide
- hydrogen peroxide
- sodium perborate

Inactive ingredients:

- thickening agents
- carrier
- surfactant
- aroma pigment
- preservative agents

hydrogen peroxide is unstable. Hydrogen peroxide penetrates the dentinal tubules and causes the formation of reactive oxygen molecules and free radicals. These released molecules break down the organic molecules that cause tooth discoloration into small pieces, allowing them to diffuse to the outer surface of the tooth and making the tooth appear lighter by reflecting the light of its own color (Bizhang, Domin, Danesh, and Zimmer 2017; Nezir and Özcan, 2023).

Carbamide peroxide interacts with water, causing the formation of hydrogen peroxide. It has a whitening mechanism similar to hydrogen peroxide (Carey 2014).

Hydrogen peroxide is released during the decomposition of sodium perborate. The released hydrogen peroxide provides the whitening process by producing different radicals or ions depending on the pH value, light effect, temperature, and the presence of co-catalysts and metallic reactions.

4.4.6 Side Effects of the Whitening Process on Teeth

Although whitening procedures are considered safe, whitening agents have some side effects.

These are as follows:

i. tooth sensitivity
ii. resorptions in dental hard tissues and damage to the pulp
iii. allergic reactions
iv. toxicity
v. effects on bond strength of resin composites (Nezir and Özcan, 2023).

Considering these side effects, studies on alternative whitening agents have increased recently. In these studies, tooth whitening effectiveness was evaluated using UV-VIS spectroscopy.

In a study by Kamaraj and Raguganesh, they prepared a new whitening solution containing phosphate group bonds, taking into account the chelation effect and whitening effect of ultraphosphate (Kamaraj and Raguganesh 2020). They compared this newly prepared solution with 30% hydrogen peroxide and another solution known as McInnes solution (McInnes solution occurred with 10 ml of 36% hydrochloric acid, 10 ml of 30% hydrogen peroxide, and 2 ml of anesthetic ether). Since ultraviolet-visible spectrophotometry is a method routinely used in analytical chemistry for the quantitative determination of different analytes such as transition metal ions, highly conjugated organic compounds, and biological macromolecules, they used this analysis to determine the absorber concentration in the solution they prepared and compare it with the other two solutions. In this study, 60 extracted teeth were kept in coffee and colored and used. The maximum absorbance value of coffee was determined by spectrophotometer. Colored tooth samples were kept in the solutions for 30 minutes, and maximum absorbance values were determined using a spectrophotometer. As a result of UV-VIS analysis, they stated that the absorbance value of coffee was 2.23, followed by ultra phosphate (1.23), hydrogen peroxide (0.89), and McInnes solution (0.67). According to these absorbance values, it was seen that the amount of caffeine in the teeth in the ultraphosphate solution was higher than in other solutions. UV-VIS analysis results in this study showed that ultra phosphate removed stains without breaking down caffeine. As a result of this study, it was emphasized that the stain removal effectiveness of ultra phosphate was comparable to the other two solutions and that it removed stains without creating a whitening effect on the teeth (Kamaraj and Raguganesh 2020).

In a study conducted by Mazilu Moldovan et al., they aimed to obtain whitening gels with superior properties such as providing stability over time, repairing damage to the enamel structure after bleaching, reducing existing pigmentation, and rapid bleaching (Mazilu Moldovan et al. 2019). For this purpose, they prepared 3 experimental whitening gels containing natural fruit juices. They examined the effects of these gels on the surface morphology and color of different composite materials. Additionally, they compared the cytotoxicity of the gels and nanocomposites after bleaching with a commercially available bleaching gel. They used UV-VIS spectroscopy, computerized measurement, and fluorescence spectrometry to observe color changes, and scanning electron microscopy (SEM) to evaluate the microstructure of the surface. They stated that they evaluated the samples with a UV-VIS spectrophotometer after 10 and 30 days to determine the color difference of the colored samples prepared with the composites. They indicated that determined reflectance spectrum recorded at 380–770 by comparison with a spectral standard. They reported the analysis results as showing that the average color difference after bleaching was significantly reduced for all bleaching gels. As a result of this study, it was concluded that experimental natural whitening gels can lead to a reduction in discoloration without any change in the surface morphology of dental composites (Mazilu Moldovan et al. 2019).

In another study, considering the possible side effects or toxic effects of peroxide-based whitening gels, alternative whitening methods based on natural compounds that have a remineralizing effect on tooth enamel were investigated (Mazilu et al. 2021). They used bromelain extract, quince extract, and whey because they are natural and have antimicrobial and physiological effects. They prepared five natural product gels with these products in various proportions. They also prepared two additional gels, one containing lubrizol and the other containing SiO_2. All gels were analyzed by Gel electrophoresis under denaturing conditions (SDS-PAGE analysis), UV-VIS and FTIR spectroscopy, and SEM microscopy in this study. Alternatively, gels were dissolved in distilled water and measured in solution using UV-VIS spectrophotometers. In the relevant study, it was stated that wavelengths varied between 200 and 700 nm. In the UV-VIS spectra, it was concluded that the 303 nm band was assigned to an in situ modified form of bromelain. When all analyzes were evaluated, it was argued that bromelain gel could be used for its therapeutic effects in oral injuries of soft tissues and as a whitening tool for tooth enamel (Mazilu et al. 2021).

It is known that hydrogen peroxide spreads into the pulp chamber and causes the pulp to lose its vitality. Mena-Serrano et al. published a study on the grounds that sodium ascorbate prevents the spread of hydrogen peroxide into the pulp chamber. In the relevant study, they aimed to evaluate the effect of 10% sodium ascorbate application after office bleaching with 35% hydrogen peroxide on the spread of hydrogen peroxide into the pulp chamber, color change, and microtensile bond strength to enamel. In this study conducted on extracted teeth, hydrogen peroxide of known concentration was determined by measuring it with a UV-VIS spectrophotometer and served as a reference for calculating the results of the study samples. Some of the extracted teeth were designated as the control group, and no bleaching was applied. It has been stated that an acetate buffer was placed in the pulp chamber of each sample to protect and absorb all hydrogen peroxide from whitening that could reach the pulp chamber of the teeth. In this study, teeth that underwent office bleaching were divided into two groups, and 10% sodium ascorbate was applied to one group and not to the other group. The hydrogen peroxide concentration in acetate buffer was determined by UV-VIS spectrophotometer with the maximum absorbance peak at 596 nm. Statistically significant differences were found between groups in terms of hydrogen peroxide concentration. The hydrogen peroxide concentration, from lowest to highest, is indicated as control group, sodium acetate applied group, and non-treated group, respectively (Mena-Serrano et al. 2023).

4.5 EVALUATION OF THE ANTIBACTERIAL EFFICACY AND STRUCTURE OF COMPOSITE FILLINGS

4.5.1 Dental Composite

Dental composites are filling materials used to replace the loss of material in the teeth caused by caries, minimal trauma, etc. Dental composites were first produced in 1960 (Baum, Phillips, and Lund 1985). These first produced dental composites were chemically polymerized. In 1970, light-curing resins were developed, which was considered a major advance in dental composites (Baum, Phillips, and Lund 1985).

Dental composite resins consist of the following:

I. organic matrix
II. inorganic filler
III. interconnecting agent (silane)
IV. color fixers
V. color pigments
VI. activation systems that provide polymerization (Phillips 1981).

4.5.1.1 Organic Matrix

The organic matrix consists of high molecular weight monomers bisphenol A glycidyl methacrylate (Bis-GMA) or urethane dimethacrylate (UDMA).

Bis-GMA are rigid polymers with terminal methacrylate groups and two benzene rings around the center. With the use of a peroxide catalyst and amine accelerator, additional polymerization and two reactive double bonds can occur (Söderholm and Mariotti 1999). The disadvantages of Bis-GMA as a dental filling material are its inability to provide color stability and its high viscosity. To avoid the disadvantage of high viscosity, diluent monomers such as low molecular weight triethyleneglycol dimethacrylate (TEGDMA) and ethyleneglycol dimethacrylate (EGDMA) were added into the organic matrix (Feilzer and Dauvillier 2003; Fróes-Salgado et al. 2012). By adding these monomers, the amount of cross-linking and hardness were increased while viscosity was reduced (Feilzer and Dauvillier 2003; Fróes-Salgado et al. 2012).

UDMA is another monomer used as an organic matrix. It has low viscosity compared to Bis-GMA. In addition to this advantage, there is a highly flexible urethane bond that can increase the toughness of resin composites based on this monomer (Barszczewska-Rybarek 2009).

4.5.1.2 Inorganic Filler

Inorganic filler particles can consist of different materials in composites produced by different companies. However, the most commonly used materials are as follows:

colloidal silica

barium silicate

strontium/borosilicate glass

quartz

zinc silicate

lithium aluminum silicate

silicon dioxide, etc. (Ferracane 2006).

Inorganic fillers not only consist of different materials but also vary in shape, size, and chemical composition. Inorganic filler determines the physical and mechanical properties of the composite. Thus, it ensures the low coefficient of thermal expansion of the composite, reduction in polymerization shrinkage, radiopacity, and strengthens the aesthetic results (Labella, Lambrechts, Van Meerbeek, and Vanherle 1999; Habib et al. 2016).

Inorganic filler particles are classified as follows according to their size:

- Macro fillers: 10–100 μm
- Midi fillers: 1–10 μm
- Mini fillers: 0.1–1 μm
- Micro fillers: 0.01–0.1 μm
- Nano fillers: 0.005–0.01 μm.

4.5.1.3 Interconnecting Agent (Silane)

Silanes, known as interconnecting agents, provide circular bonding of organic matrix and inorganic fillers to each other. This feature prevents hydrolytic cracking, which causes resin breakage at the filler and resin matrix interface and allows stress transfer between the filler and the resin matrix (Musanje and Ferracane 2004; Pu, Mark, Jethmalani, and Ford 1997). Organo adhesives, which are organosilane compounds, are most commonly used. The most preferred organosilanes is gamma-methacryl oxypropyltrimethoxysilane. This organosilane is a bifunctional molecule attached to the filler particle with its hydroxyl group and to the resin matrix during the polymerization of the composite with its methacrylate group (Wilson and Antonucci 2006).

4.5.1.4 Features Sought in Dental Composites

- In order for the composite to be placed on the tooth surface easily and to prevent marginal gaps, the volumetric shrinkage or expansion of the composite during polymerization should be low.
- In order for the process to be short, the polymerization time should also be short. Therefore, the polymerization degree of the composite must be high.
- Must adapt to oral conditions.
- It must have good mechanical and physical properties. Must be able to tolerate stress and pressure.
- It must have high color and light stability. This feature should remain aesthetic for a long time.
- It should not show toxicity or have mutagenic or carcinogenic effects.
- It should not allow secondary caries formation.
- It must have the ability to remineralize against tooth substance loss.
- Must be able to repair itself (Fugolin and Pfeifer 2017; Moszner and Salz 2001).

Many studies have been done to obtain an ideal composite material. However, making materials that meet all of their ideal properties is a very challenging process for any engineering. The ideal composite has yet to be obtained so far. Today's composites have been developed into wear-resistant, low-stress materials. Most of the research in this field focused on obtaining composites with antibacterial, remineralizing, or bioactive/biomimetic properties. One of the major limitations of today's dental composite resins is polymerization shrinkage. Polymerization shrinkage forms a gap between the tooth and the composite, and microleakage of bacteria that causes secondary caries occurs from this gap (Kruly et al. 2018; Veloso et al. 2019; Meereis et al. 2018). New therapeutic composites with antibacterial properties have been developed to prevent secondary caries formation (Sun et al. 2021). As antibacterial agents, chlorhexidine, chitosan, silver, and metallic nanoparticles were added to the composites (Sun et al. 2021).

In a study by Arif et al., they synthesized a new type of antibacterial resin composite containing ciprofloxacin-loaded silver nanoparticles (CIP-AgNP) (Arif et al. 2022). They evaluated the new composite they synthesized using Fourier transform infrared spectrometer (FTIR), zeta potential, SEM, and UV-VIS spectroscopy, based on silver nanoparticles (AgNP) (Arif et al. 2022). In UV-VIS analysis for AgNPs and CIP-AgNPs, it was observed that they peaked in the 400–420 nm range (Arif et al. 2022). While AgNPs had a strong and broad peak at 404 nm, it was concluded that after ciprofloxacin loading, this peak became a strong and broad peak at 414 nm (Arif et al. 2022). Together with other analyses, it was concluded that adding CIP-AgNPs improved dental resin composites' mechanical and antibacterial activity (Arif et al. 2022).

In a study conducted by Tanweer et al., they aimed to improve the antibacterial potential of dental composite resin with magnesium-doped zinc oxide (Mg-ZnO) nanoparticles (Nps), considering that secondary caries that may occur under dental composite restorations are more common in people with diabetes and chronic diseases (Tanweer et al. 2022). They preferred Mg and ZnO due to their antibacterial properties. In this study, multiple characterization analyzes were performed using UV-VIS spectrophotometry, scanning electron microscopy (SEM) and energy dispersive X-ray spectroscopy analyses, and Nova NanoSEM 450 field emission scanning electron microscope to confirm the synthesis of ZnO and Mg-ZnO NPs. The UV peak of ZnO NPs was obtained at 362 nm, confirming its synthesis. Its synthesis was confirmed by showing a shift to 358 nm after integration with Mg. As a result of the analyses, it was stated that Mg-ZnO NPs improved the antibacterial properties of resin composites without compromising mechanical properties, aesthetics, and biocompatibility, did not allow secondary caries formation by preventing biofilm formation with their pH buffering effect, and could be an effective candidate for caries treatment even in high-risk diabetic patients.

Another recent study is based on the hypothesis that the disadvantages of dental composites, such as relatively high brittleness and low strength, can be eliminated by coating NPs and NPs with polymer (Ok 2022). In this study, silica nanoparticles (Si-NPs), ZnO NPs, and their silica/polymethylmethacrylate (Si/PMMA) and zinc oxide/chitosan (ZnO/CS) core-shell nanostructures were successfully synthesized and characterized. Dynamic light scattering and zeta potential, FTIR spectrometer, SEM, UV-VIS spectrophotometer, thermogravimetric analysis, and X-ray diffraction were used for characterization. The absorbance of NPs was measured in the 280–600 nm region by UV–VIS. According to the UV-VIS spectrum, the absorption peaks of Si-based NPs were determined as 286 nm. While the absorption peaks of ZnO NPs were at 358 nm, the absorption peaks of ZnO/CS core-shell NPs shifted to 363 nm. According to the analysis results, it was stated that the dental composite containing 0.1% by weight Si/PMMA and 0.5% by weight ZnO/CS had both improved mechanical properties and an improved antibacterial effect (Ok 2022).

4.6 EVALUATION OF ALTERNATIVE HERBAL EXTRACTS TO PREVENT DENTAL CARIES AND PERIODONTITIS

4.6.1 Oral Infections

Oral infections are a past and present public health problem. Oral infections include dental caries, periodontitis, pulpal diseases, apical diseases, peri-implant diseases, candidiasis, and many other oral infections, but the most common ones globally are dental caries and periodontitis (O Ahmed, Sibuyi, Fadaka, Madiehe, Maboza, Meyer, and Geerts 2022a).

4.6.1.1 Dental Caries

Tooth decay is a condition in which some bacteria in the mouth, especially Streptococcus mutans, soften (demineralization) the hard tooth structure by causing ion losses in the tooth structure in the presence of certain factors. The factors necessary for the formation of caries are:

presence of cariogenic (cavity-causing) bacteria in the mouth

food (sugar) needed for these bacteria to feed

enzymes released by bacterial nutrition or acid formed by consuming some foods (therefore low pH) the time required for the acid to act on the hard tissues of the tooth (Pitts, Twetman, Fisher, and Marsh 2021).

4.6.1.2 Periodontitis

Periodontitis is a chronic oral infection characterized by the progressive destruction of the periodontal ligament and alveolar bone, which are tooth-supporting tissues, due to various factors, especially specific bacteria, destructive responses of the host immune system, and smoking (Kwon, Lamster, and Levin 2021).

Dental plaque is the main cause of tooth decay and periodontitis (Kwon, Lamster, and Levin 2021; O. Ahmed, Sibuyi, Fadaka, Madiehe, Maboza, Olivier, et al. 2022b). Dental plaque is a layer containing many bacteria, formed by the accumulation of unremoved food residues on the teeth. Therefore, maintaining personal oral hygiene comes to the fore to prevent the formation of caries and periodontitis. Toothbrushes, mouthwashes, and toothpaste are used for personal oral hygiene.

Recently, due to their antimicrobial, anti-inflammatory, antifungal, and antioxidant properties, silver nanoparticles (AgNPs) have been incorporated into products used for personal oral hygiene to help prevent oral infections. In a study conducted by Ahmed et al., they examined the antimicrobial effects of AgNPs synthesized using Gum Arabic extracts (GAE) (O. Ahmed, Sibuyi, Fadaka, Madiehe, Maboza, Olivier, et al. 2022b). They stated that they used UV-VIS spectrophotometry, dynamic light scattering (DLS), transmission electron microscopy (TEM), and FTIR spectroscopy to synthesize and characterize GA-AgNPs. They used GAE and two $AgNO_3$ at concentrations of 0.4 g and 0.1 g to produce GA-AgNPs. They obtained GA-AgNPs_0.1 g and GA-AgNPs_0.4 g. They measured at 300–650 nm using the UV-VIS spectrum. They stated that the presence of GA-AgNPs was confirmed by UV-VIS spectra analysis. As a result of this analysis, they stated that a UV-VIS spectrum with surface plasmon resonance (SPR) and maximum absorbance (λmax) around 400 nm is characteristic for AgNPs. They found the SPR values to be 450 and 425 nm for GA-AgNPs-0.1 g and GA-AgNPs-0.4 g, respectively. They found that the peak intensity of GA-AgNPs synthesized with 0.1 g $AgNO_3$ was slightly lower and wider than those synthesized with 0.4 g $AgNO_3$. Based on this analysis result, they stated that GA-AgNPs_0.1 g was polydispersic, and more AgNPs were formed at 0.4 g concentration. Together with other analysis results, they reported that GA-AgNPs can be used as additives to dental products because they can bind to enamel and prevent bacterial biofilm formation on teeth (O. Ahmed, Sibuyi, Fadaka, Madiehe, Maboza, Olivier, et al. 2022b).

Another similar study aimed to evaluate whether the antimicrobial activity of GA-AgNPs could be modified when included in a toothpaste formulation (O.A.K. Ahmed et al. 2023). In the relevant study, a total of four commercial toothpastes were used: three fluoride-based and one coal-based (O.A.K. Ahmed et al. 2023). GA-AgNPs_0.4 g was synthesized, and UV-VIS, DLS, TEM, and FTIR spectrophotometer were used for characterization analysis (O.A.K. Ahmed et al. 2023). As a result of all analyses, the antimicrobial activity of GA-AgNPs_0.4 g did not change after combining with a fluoride-based toothpaste. It was stated that cytotoxicity varies depending on concentration and time of use, and the need for more meticulous studies on this subject was emphasized (O.A.K. Ahmed et al. 2023).

4.7 ASSESSMENT OF THE OPACITY OF MATERIALS USED IN DENTAL PROSTHESES

Fixed or removable dental prostheses are made to restore aesthetics, function, and phonation in cases of excessive tooth loss or missing teeth.

Fixed dentures can be applied to teeth where tooth tissue loss is high and/or where tooth deficiencies are few and there are adjacent teeth that can provide support in the mouth and absorb the chewing force on the missing teeth in order to replace the missing teeth. Various materials (gold, porcelain, etc.) have been used to make fixed prostheses until today. Recently, dental ceramics have been used due to the demands that prioritize aesthetic developments. The most preferred material for dental ceramics is zirconium (Sailer et al. 2018; Kern, Passia, Sasse, and Yazigi 2017).

Zirconium ceramic has a polymorphic structure and can be found in three crystal structures: monoclinic, tetragonal, and cubic. To be used in a clinical environment, these crystal structures must be stable at room temperature and have sufficient mechanical and optical properties. Pure zirconium, which is in the monoclinic phase at room temperature, transitions to the tetragonal phase above 1170 °C. The tetragonal phase transitions to the cubic phase at temperatures much higher than room temperature (Piconi and Maccauro 1999; Raigrodski 2004). Stabilizer oxides (CaO, MgO, CeO_2, Y_2O_3) are used to ensure the stability of zirconium at room temperature (Piconi and Maccauro 1999). Yttrium oxide (Y_2O_3) is the most used stabilizer due to its excellent luminescence properties such as color purity and chemical and thermal stabilities (Lughi and

Sergo 2010; Michailova et al. 2020). Traditional white opaque zirconium was obtained by sintering with 3 mol% Y_2O_3. It has been developed by applying various modifications to enable the use of traditional zirconia as monolithic restorations with adequate aesthetics. Therefore, the yttrium content was increased from 3 mol% (3Y-TZP) to 4 mol% (4Y-TZP) and 5 mol% (5Y-TZP) for higher translucency (Stawarczyk et al. 2017). In 2015, polychromatic (multilayered) zirconiums were introduced to the market to imitate the color of enamel and dentin (Kaizer et al. 2020). To date, new-generation zirconium ceramics have been developed and continue to be developed.

In a recent study, fracture load, two-body wear behavior, and translucency were examined to evaluate the mechanical properties of new generation zirconium (Michailova et al. 2020). Thirty-six samples were created using three different commercially available multilayer zirconium materials and lithium disilicate ceramics. UV-VIS was used to evaluate the translucency of the samples. Quantitative measures of translucency were determined by analyzing the light that could be transmitted in each sample. In this study, the overall light transmittance for each sample was obtained by calculating the integration of transmission coefficients at wavelengths between 400 and 700 nm. Light transmission analysis was obtained as a percentage by dividing the overall light transmittance by the light transmittance obtained without the sample. As a result, a statistically significant difference was found between enamel and dentin layers in each zirconium material. It has also been stated that although zirconium has higher mechanical properties than lithium disilicate ceramic, its translucency is less.(Michailova et al. 2020)

The quality of zirconium ceramic restorations depends on the bonding effectiveness of the zirconium surface and adhesive system (Moon et al. 2016). Today, the bonding effect of zirconium ceramic restorations to the teeth or implant surface is achieved by dentists by sandblasting with alumina particles and increasing the surface roughness (Moon et al. 2016). With this technique, the material is abraded by hitting the surface with hard particles at high speeds, and a rough surface is obtained. In addition, this technique allows for cleaning the ceramic surface, removing impurities, increasing the surface area, and changing the surface energy and wettability (Moon et al. 2016). Thus, micromechanical coupling is achieved. However, this sandblasting technique damages the mechanical properties of zirconium ceramics (Moon et al. 2016).

UV-VIS was used for optical characterization in the study to investigate how the sandblasting technique with alumina particles affects the optical properties as well as the mechanical properties (Schabbach et al. 2023). Samples were obtained from four different layers of multilayer zirconium, and these samples were sandblasted with alumina particles (Schabbach et al. 2023). Reflectance measurements were carried out on white and black backgrounds, and UV-visible light wavelengths under both backgrounds were shown as 400–700 nm (Schabbach et al. 2023). The colorimetric parameters of the CIELab system of the samples were obtained separately on white and black backgrounds (Schabbach et al. 2023). The translucency parameter was obtained using Equal 1 (Schabbach et al. 2023). According to the results obtained, it was reported that the sandblasting process did not affect the optical properties of multilayer translucent zirconia (Schabbach et al. 2023).

4.8 CONCLUSION

UV-VIS spectroscopy is used in the research and development of materials used in dentistry. In this regard, this method is an analysis method that can contribute to the development of new production studies and technologies such as ensuring the longevity of dental materials, their aesthetics, and the absence of harmful effects on health.

REFERENCES

Abdolkarimi-Mahabadi, M., A. Bayat, and A. Mohammadi. 2021. "Use of UV-Vis spectrophotometry for characterization of carbon nanostructures: A review." *Theoretical and Experimental Chemistry* 57 (3): 191–198. 10.1007/s11237-021-09687-1

Ahmed, O., N.R.S. Sibuyi, A.O. Fadaka, M.A. Madiehe, E. Maboza, M. Meyer, and G. Geerts. 2022a. Plant extract-synthesized silver nanoparticles for application in dental therapy. *Pharmaceutics*. 14 (2). 380.

Ahmed, O. A. K., N. R. S. Sibuyi, A. O. Fadaka, E. Maboza, A. Olivier, A. M. Madiehe, M. Meyer, and G. Geerts. 2023. "Prospects of using gum arabic silver nanoparticles in toothpaste to prevent dental caries." *Pharmaceutics* 15 (3). 10.3390/pharmaceutics15030871

Ahmed, O., N. R. S. Sibuyi, A. O. Fadaka, A. M. Madiehe, E. Maboza, A. Olivier, M. Meyer, and G. Geerts. 2022b. "Antimicrobial effects of gum arabic-silver nanoparticles against oral pathogens." *Bioinorganic Chemistry and Applications* 2022: 9602325. 10.1155/2022/9602325

Alqahtani, Mohammed Q. 2014. "Tooth-bleaching procedures and their controversial effects: A literature review." *The Saudi Dental Journal* 26 (2): 33–46.

Arens, Donald. 1989. "The role of bleaching in esthetics." *Dental Clinics of North America* 33 (2): 319–336.

Arif, W., N. F. Rana, I. Saleem, T. Tanweer, M. J. Khan, S. A. Alshareef, H. M. Sheikh, F. S. Alaryani, M. O. Al-Kattan, H. A. Alatawi, F. Menaa, and A. Y. Nadeem. 2022. "Antibacterial activity of dental composite with ciprofloxacin loaded silver nanoparticles." *Molecules* 27 (21). 10.3390/molecules27217182

Attin, Thomas, F Paque, F Ajam, and AM Lennon. 2003. "Review of the current status of tooth whitening with the walking bleach technique." *International Endodontic Journal* 36 (5): 313–329.

Barszczewska-Rybarek, Izabela M. 2009. "Structure–property relationships in dimethacrylate networks based on Bis-GMA, UDMA and TEGDMA." *Dental Materials* 25 (9): 1082–1089.

Başar, Ö. and Şanal Çıkman, A. 2023. Dental Bleaching Procedures. In: Hatipoğlu, Ö. (ed.), Current Dentistry Research- I. Özgür Publications. 10.58830/ozgur.pub127.c580

Baum, L, RW Phillips, and MR Lund. 1985. Textbook of operative dentistry. *Tooth Colored Restoratives* (2 bs.). Philadelphia: Saunders.

Bizhang, Mozhgan, Julia Domin, Gholamreza Danesh, and Stefan Zimmer. 2017. "Effectiveness of a new non-hydrogen peroxide bleaching agent after single use-a double-blind placebo-controlled short-term study." *Journal of Applied Oral Science* 25: 575–584.

Carey, C. M. 2014. "Tooth whitening: What we now know." *The Journal of Evidence-Based Dental Practice* 14 Suppl: 70–76. 10.1016/j.jebdp.2014.02.006

Dahl, JE, and U Pallesen. 2003. "Tooth bleaching—a critical review of the biological aspects." *Critical Reviews in Oral Biology & Medicine* 14 (4): 292–304.

Diffey, Brian L. 2002. "Sources and measurement of ultraviolet radiation." *Methods* 28 (1): 4–13.

Duckworth, James H. 1998. "Quantitative analysis." *Applied Spectroscopy: A Compact Reference for Practitioners* 93: 29–47.

Feilzer, AJ, and BS Dauvillier. 2003. "Effect of TEGDMA/BisGMA ratio on stress development and viscoelastic properties of experimental two-paste composites." *Journal of Dental Research* 82 (10): 824–828.

Ferracane, Jack L. 2006. "Is the wear of dental composites still a clinical concern?: is there still a need for in vitro wear simulating devices?" *Dental Materials* 22 (8): 689–692.

Fróes-Salgado, NRG, LC Boaro, B Pick, CS Pfeifer, Carlos Eduardo Francci, MM Méier, and Roberto Ruggiero Braga. 2012. "Influence of the base and diluent methacrylate monomers on the polymerization stress and its determinants." *Journal of Applied Polymer Science* 123 (5): 2985–2991.

Frysh, H., W. H. Bowles, F. Baker, F. Rivera-Hidalgo, and G. Guillen. 1995. "Effect of pH on hydrogen peroxide bleaching agents." *Journal of Esthetic Dentistry* 7 (3): 130–133. 10.1111/j.1708-8240.1995.tb00730.x

Fugolin, APP, and CS Pfeifer. 2017. "New resins for dental composites." *Journal of Dental Research* 96 (10): 1085–1091.

Greenwall-Cohen, Joseph, and Linda H Greenwall. 2019. "The single discoloured tooth: vital and non-vital bleaching techniques." *British Dental Journal* 226 (11): 839–849.

Habib, Eric, Ruili Wang, Yazi Wang, Meifang Zhu, and XX Zhu. 2016. "Inorganic fillers for dental resin composites: present and future." *ACS Biomaterials Science & Engineering* 2 (1): 1–11.

Harlan, AW. 1884. "The removal of stains from the teeth caused by the administration of medicinal agents and bleaching of pulpless teeth." *Welcome to the American Journal of Dentistry* 18: 521–524.

Haywood, Van B, and Harald O Heymann. 1989. "Nightguard vital bleaching." *Quintessence International* 20 (3): 173–176.

Kaczmarek, K., A. Leniart, B. Lapinska, S. Skrzypek, and M. Lukomska-Szymanska. 2021. "Selected spectroscopic techniques for surface analysis of dental materials: A narrative review." *Materials (Basel)* 14 (10): 2624. 10.3390/ma14102624

Kaizer, Marina R, Nantawan Kolakarnprasert, Camila Rodrigues, Herzl Chai, and Yu Zhang. 2020. "Probing the interfacial strength of novel multi-layer zirconias." *Dental Materials* 36 (1): 60–67.

Kamaraj, Palani Selvi, and Vidhya Raguganesh. 2020. "Ultraviolet–visible spectrophotometric analysis of tooth whitening efficiency of ultraphosphate." *Journal of Contemporary Medicine and Dentistry* 8 (1): 48–53.

Kern, Matthias, Nicole Passia, Martin Sasse, and Christine Yazigi. 2017. "Ten-year outcome of zirconia ceramic cantilever resin-bonded fixed dental prostheses and the influence of the reasons for missing incisors." *Journal of Dentistry* 65: 51–55.

Kruly, Paula de Castro, Marcelo Giannini, Renata Correa Pascotto, Laise Midori Tokubo, Uhana Seifert Guimarães Suga, Any de Castro Ruiz Marques, and Raquel Sano Suga Terada. 2018. "Meta-analysis of the clinical behavior of posterior direct resin restorations: Low polymerization shrinkage resin in comparison to methacrylate composite resin." *PLoS One* 13 (2): e0191942.

Kury, M., R. B. E. Lins, B. A. Resende, M. Z. D. Picolo, C. B. André, and V. Cavalli. 2022. "The influence of the renewal or the single application of the peroxide gel on the efficacy and tooth sensitivity outcomes of in-office bleaching-A systematic review and meta-analysis." *Journal of Esthetic and Restorative Dentistry* 34 (3): 490–502. 10.1111/jerd.12827

Kwon, TaeHyun, Ira B Lamster, and Liran Levin. 2021. "Current concepts in the management of periodontitis." *International Dental Journal* 71 (6): 462–476.

Labella, R, Paul Lambrechts, Bart Van Meerbeek, and Guido Vanherle. 1999. "Polymerization shrinkage and elasticity of flowable composites and filled adhesives." *Dental Materials* 15 (2): 128–137.

Lughi, Vanni, and Valter Sergo. 2010. "Low temperature degradation-aging-of zirconia: A critical review of the relevant aspects in dentistry." *Dental Materials* 26 (8): 807–820.

Maran, Bianca Medeiros, Adrieli Burey, Thalita de Paris Matos, Alessandro D Loguercio, and Alessandra Reis. 2018. "In-office dental bleaching with light vs. without light: A systematic review and meta-analysis." *Journal of Dentistry* 70: 1–13.

Mazilu, A., V. Popescu, C. Sarosi, R. S. Dumitrescu, A. M. Chisnoiu, M. Moldovan, L. S. Dumitrescu, D. Prodan, R. Carpa, G. F. Gheorghe, and R. M. Chisnoiu. 2021. "Preparation and in vitro characterization of gels based on bromelain, whey and quince extract." *Gels* 7 (4). 10.3390/gels7040191

Mazilu Moldovan, A., C. Sarosi, M. Moldovan, F. Miuta, D. Prodan, A. Antoniac, C. Prejmerean, L. Silaghi Dumitrescu, V. Popescu, A. D. Raiciu, and V. Saceleanu. 2019. "Preparation and characterization of natural bleaching gels used in cosmetic dentistry." *Materials (Basel)* 12 (13). 10.3390/ma12132106

Meereis, Carine Tais Welter, Eliseu Aldrighi Münchow, Wellington Luiz De Oliveira Da Rosa, Adriana Fernandes Da Silva, and Evandro Piva. 2018. "Polymerization shrinkage stress of resin-based dental materials: A systematic review and meta-analyses of composition strategies." *Journal of the Mechanical Behavior of Biomedical Materials* 82: 268–281.

Mena-Serrano, A., M. G. Granda-Albuja, J. Naranjo, E. A. Fierro, M. W. Favoreto, A. D. Loguercio, and A. Reis. 2023. "Effects of the application of sodium ascorbate after in-office bleaching on the penetration of hydrogen peroxide, color change, and microtensile bond strength." *Brazilian Dental Journal* 34 (5): 87–94. 10.1590/0103-6440202305214

Michailova, M., A. Elsayed, G. Fabel, D. Edelhoff, I. M. Zylla, and B. Stawarczyk. 2020. "Comparison between novel strength-gradient and color-gradient multilayered zirconia using conventional and high-speed sintering." *Journal of the Mechanical Behavior of Biomedical Materials* 111: 103977. 10.1016/j.jmbbm.2020.103977

Moon, Ji-Eun, Sung-Hun Kim, Jai-Bong Lee, Jung-Suk Han, In-Sung Yeo, and Seung-Ryong Ha. 2016. "Effects of airborne-particle abrasion protocol choice on the surface characteristics of monolithic zirconia materials and the shear bond strength of resin cement." *Ceramics International* 42 (1): 1552–1562.

Moszner, Norbert, and Ulrich Salz. 2001. "New developments of polymeric dental composites." *Progress in Polymer Science* 26 (4): 535–576.

Musanje, Lawrence, and Jack L Ferracane. 2004. "Effects of resin formulation and nanofiller surface treatment on the properties of experimental hybrid resin composite." *Biomaterials* 25 (18): 4065–4071.

Nezir, Merve, and Suat Özcan. 2023. "Diş Ağartma Tedavilerine Güncel Bir Bakış." *Selcuk Dental Journal* 10 (1): 112–117.

Ok, İzel. 2022. "Production of Dental Filling Composite Material with Improved Mechanical and Antibacterial Properties by ZnO/Chitosan Doped Silica/PMMA Core/Shell Nano Hybrid Structured Inorganic Matrix." Order No. 31064292, Izmir Katip Celebi University (Turkey). https://www.proquest.com/dissertations-theses/production-dental-filling-composite-material-with/docview/2918170075/se-2

Oktay, EK. 2006. "Farklı vital beyazlatma sistemlerinin diş rengi üzerine etkilerinin klinik olarak karşılaştırılması." Doktora Tezi (Thesis), Hacettepe Üniversitesi Sağlık Bilimleri Enstitüsü. *Ankara, Türkiye.* https://tez.yok.gov.tr/UlusalTezMerkezi/tezDetay.jsp?id=pEk46agRzX08kG80fCUsTA&no=fw7gAQBO7vwxM_SkT9mktQ

Pentassuglia, Simona, Valeria Agostino, and Tonia Tommasi. 2018. "EAB—electroactive biofilm: a biotechnological resource." In *Reference Module in Chemistry, Molecular Sciences and Chemical Engineering*, 110–123. Elsevier.

Phillips, Ralph W. 1981. "Past, present, and future composite resin systems." *Dental Clinics of North America* 25 (2): 209–218.

Piconi, Corrado, and G Maccauro. 1999. "Zirconia as a ceramic biomaterial." *Biomaterials* 20 (1): 1–25.

Pitts, N. B., S. Twetman, J. Fisher, and P. D. Marsh. 2021. "Understanding dental caries as a non-communicable disease." *British Dental Journal* 231 (12): 749–753. 10.1038/s41415-021-3775-4

Plotino, G., L. Buono, N. M. Grande, C. H. Pameijer, and F. Somma. 2008. "Nonvital tooth bleaching: a review of the literature and clinical procedures." *Journal of Endodontia* 34 (4): 394–407. 10.1016/j.joen.2007.12.020

Pu, Zhengcai, James E Mark, Jagdish M Jethmalani, and Warren T Ford. 1997. "Effects of dispersion and aggregation of silica in the reinforcement of poly (methyl acrylate) elastomers." *Chemistry of Materials* 9 (11): 2442–2447.

Punia Sandhya, Kapoor, Arora Vipin. 2015. "Chapter 26: Management of discolored teeth." In *Textbook of Operative Dentistry*, edited by Amit Garg Nisha Garg, 447–462. The Health Sciences.

Raigrodski, Ariel J. 2004. "Contemporary all-ceramic fixed partial dentures: a review." *Dental Clinics* 48 (2): 531–544.

Reitzer, François, Claire Ehlinger, and Maryline Minoux. 2019. "A modified inside/outside bleaching technique for nonvital discolored teeth: a case report." *Quintessence International* 50 (10): 802–807.

Rotstein, Ilan, and Yiming Li. 2008. "Tooth discoloration and bleaching." *Endodontics* 5: 845–860.

Sailer, Irena, Marc Balmer, Jürg Hüsler, Christoph Hans Franz Hämmerle, Sarah Känel, and Daniel Stefan Thoma. 2018. "10-year randomized trial (RCT) of zirconia-ceramic and metal-ceramic fixed dental prostheses." *Journal of Dentistry* 76: 32–39.

Schabbach, L. M., B. C. Dos Santos, L. S. De Bortoli, D. Fabris, M. C. Fredel, and B. Henriques. 2023. "Translucent multi-layered zirconia: Sandblasting effect on optical and mechanical properties." *Dental Materials* 39 (9): 807–819. 10.1016/j.dental.2023.07.005

Söderholm, Karl-Johan, and Angelo Mariotti. 1999. "BIS-GMA–based resins in dentistry: are they safe?" *The Journal of the American Dental Association* 130 (2): 201–209.

Spasser, Herbert F. 1961. "A simple bleaching technique using sodium perborate." *The New York State Dental Journal* 27: 332–334.

Stawarczyk, Bogna, Christine Keul, Marlis Eichberger, David Figge, Daniel Edelhoff, and Nina Lümkemann. 2017. "Three generations of zirconia: From veneered to monolithic. Part I." *Quintessence International* 48 (5): 369–380.

Sun, Qiannan, Lingyun Zhang, Rushui Bai, Zimeng Zhuang, Yunfan Zhang, Tingting Yu, Liying Peng, Tianyi Xin, Si Chen, and Bing Han. 2021. "Recent progress in antimicrobial strategies for resin-based restoratives." *Polymers* 13 (10): 1590.

Tanweer, T., N. F. Rana, I. Saleem, I. Shafique, S. M. Alshahrani, H. A. Almukhlifi, A. S. Alotaibi, S. A. Alshareef, and F. Menaa. 2022. "Dental composites with magnesium doped zinc oxide nanoparticles prevent secondary caries in the alloxan-induced diabetic model." *International Journal of Molecular Sciences* 23 (24). 10.3390/ijms232415926

Ten Bosch, JJ, and JC Coops. 1995. "Tooth color and reflectance as related to light scattering and enamel hardness." *Journal of Dental Research* 74 (1): 374–380.

Tessier, Jerónimo, Patricia N Rodriguez, Fima Lifshitz, Silvia M Friedman, and Eduardo J Lanata. 2010. "The use of ozone to lighten teeth: An experimental study." *Acta Odontologica Latinoamericana* 23 (2): 84–89.

Truman, J. 1864. "Bleaching of non-vital discoloured anterior teeth." *Dent Times* 1 (1): 69–72.

Tüzel, Seden, and Semiha Hülya Erten Can. 2022. "Vital Diş Beyazlatma Teknikleri Ve Güncel Yaklaşımlar." *Current Research in Dental Sciences* 32 (3): 239–249.

Usta, Sıla Nur, and Eda Gözükara. 2023. "Devital Beyazlatma." *Sağlık & Bilim 2022: Medikal Araştırmalar-IV* edited by Doc. Dr. Onur Karaman, 35. Efe Akademi Yayınları İstanbul Türkiye.

Veloso, Sirley Raiane Mamede, Cleidiel Aparecido Araújo Lemos, Sandra Lúcia Dantas De Moraes, Belmiro Cavalcanti Do Egito Vasconcelos, Eduardo Piza Pellizzer, and Gabriela Queiroz De Melo Monteiro. 2019. "Clinical performance of bulk-fill and conventional resin composite restorations in posterior teeth: a systematic review and meta-analysis." *Clinical Oral Investigations* 23: 221–233.

Wang, Huaiyu, and Paul K Chu. 2013. "Surface characterization of biomaterials." In *Characterization of Biomaterials*, 105–174. Elsevier.

Wilson, Kristen S, and Joseph M Antonucci. 2006. "Interphase structure–property relationships in thermoset dimethacrylate nanocomposites." *Dental Materials* 22 (11): 995–1001.

Zimmerli, Brigitte, Franziska Jeger, and Adrian Lussi. 2010. "Bleaching of non-vital teeth." *Schweizer Monatsschrift für Zahnmedizin* 120 (4): 306–313.

CHAPTER 5

Application of Spectrometric and Chemometrics for Quality Control of Dental Materials

Tatang T. Irianti, Gunawan Indrayanto, and Abdul Rohman

5.1 INTRODUCTION

According to ISO 1942: 2020 [1], dental material can be defined as a substance or combination of substances specially formulated and prepared for use in the practice of dentistry and/or associated procedures, while dentistry is a science of preventing, diagnosing, and treating diseases and malfunctions of, and injuries to, the teeth, mouth, and jaws, and of replacing lost teeth and associated tissues and promoting oral health. In the USA and Europe, dental materials are classified as medical devices [2]. According to the Medical Device Regulation (MDR) of the European Union (EU), a medical device can be defined as any instrument, apparatus, appliance, material, or other article that did not achieve its principal intended action in or on the human body by pharmacological, immunological, or metabolic means [3, 4]. The Federation of the European Dental Industry (FIDE) summarized the classification of dental medical devices

DOI: 10.1201/9781003506805-5

according to Regulation on Medical Devices European Union 2017/745 (MDR) Annex VIII, which came into effect in the EU in May 2021 [5]. Mohn and Zehnder [6] recently reviewed and discussed in detail the impact of the MDR regulations from a dental perspective.

Schmalz and Reichl [2] published a nice review of the safety regulation and standardization of dental materials according to the ISO (14791, 10993-1, 10993-2 to 23, 7405) and their relation to the EU MDR. ISO 14791 specifies the application of risk management to medical devices. ISO 10993-1 to 23 describes the biological evaluation, chemical characterization, and analysis of dental materials. ISO 7405 specifies the evaluation of biocompatibility for medical devices, and according to this guideline, three groups of tests are described as follows: Group 1: cytotoxicity tests, including the dentin barrier test (see below); reference is also made to ISO 10993-5 (tests for in vitro cytotoxicity); Group 2: systemic toxicity, skin irritation, and intracutaneous reactivity, delayed skin sensitization, genotoxicity, and implantation tests; here no procedures are described in ISO 7405, but reference is made to the relevant ISO 10993 standards; and Group 3: dentally specific usage tests. ISO 4049 defines the standard requirement for dental polymer-based filling and restorative materials and polymer-based luting materials supplied in a form suitable for mechanical mixing, hand mixing, or intra-oral and extra-oral external energy activation and intended for use primarily for the direct or indirect restoration of cavities in the teeth [7]. Chemicals used to produce dental materials must comply with REACH (Registration, Evaluation, Authorization, and Restriction of Chemicals), which is a regulation of the EU, implemented to improve the protection of human health and the environment from the risks that can be posed by chemicals. However, under REACH, there is no obligation to register substances in amounts below 1 ton/year [8]. The OECD (Organization for Economic Co-operation and Development) provides a set of guidelines for testing chemicals, including those used in dental materials. These guidelines aim to ensure safety, assess risks, and promote best practices in chemical testing [9].

The aim of establishing international standards (e.g., ISO standards) is to issue test methods for certain products (dental materials) and thus determine fulfilment or not of related requirements, primarily for performance and safety afore on the chemical and/or physical properties with attention towards clinical relevance to contribute to improving oral and general health. From the biological perspective, standards might not contain rigid (i.e., pass/fail) requirements, but the outcome of such standardized tests is presumed to be used within a clinical risk assessment [10]. Testing of dental materials comprised unspecific tests (ISO 10993 and ISO 7405 series) and specific tests (ISO 7405) [2].

Unspecific tests do not consider the characteristics of oral cavity anatomy and physiology conditions, whereas specific tests ("usage tests") do. Unspecific tests are cytotoxicity, genotoxicity/mutagenicity, or systemic toxicity; these tests are described in the ISO 10993 series, in part of ISO 7405 and OECD guidelines. These tests are also an important part of demonstrating conformity with the general safety requirements of the MDR [2]. Dental-specific tests are comprised of (1) pulp/dentin and pulp capping usage tests, (2) endodontic usage tests, and (3) endosseous dental implant usage tests. These are mainly animal-based tests included in the ISO 7405 [2]. The main purpose of establishing such international standards as described above (ISO, MDR, REACH, OECD) is to provide test methods for the dental materials and thus ascertain fulfilment or not of related requirements, primarily for performance and safety "rather than on chemical and/or physical properties with consideration towards clinical relevance for improving oral and general health [2]. Standards play a key role in ensuring quality and safety; they should not be mistaken for comprehensive research tools. Researchers must exercise discernment when selecting methods and recognize that true scientific inquiry goes beyond mere adherence to standards [11]. The standards safeguard patient health, maintain quality, and foster transparency. Dentists, manufacturers, and regulatory bodies collaborate to uphold these standards for the benefit of dental patients. Quality control is essential to ensure dental materials meet safety standards, contribute to treatment success, and enhance patient well-being, so it should be performed using state-of-the-art methods and validated according to recent official guidelines before routine applications.

The dental care requires materials that are safe, effective, and of high quality; therefore, a detection instrument is needed urgently to determine the criteria for the utilization of these dental materials in detail and comprehensively. The characterization of the dental materials plays an important role for the quality assurance of these products so that the physical and chemical properties have to be understood deeply to confirm safety. For this task, some analytical methods are required for quality controls of dental materials. The chemical dental material properties or microscopic tests can be performed by an optical microscope and the visible light and lenses of the optical microscopes. Their signals can be described as the required information. Recently, the investigation of the microscopic tests can be carried out by fluorescence or ultraviolet-visible spectroscopy. These two instruments can be measured clinically on a patient (*in vivo*) or in the laboratory (*in vitro*). The spectroscopy detects rapidly and simply the structural changes in biological macromolecules and the visualization of particles in cells. In addition, the three-dimensional measurements can be performed with these advanced techniques. These characteristics of materials can be described better by the composition, the chemical structure, and the bonding molecule [12].

5.2 STANDARDIZATION AND QUALITY CONTROL OF DENTAL MATERIALS

Methods used for QC of dental materials comprised mechanical tests (physical) chemical analysis, and biological assays/tests. Mechanical tests used for the QC of dental materials include flexure strength, fracture toughness, and hardness. These tests play a crucial role in ensuring the reliability and performance of dental materials. The mechanical properties of dental materials directly impact the clinical performance and longevity of dental restorations [13, 14].

Some analytical methods, including physical ones such as surface analysis and chemical and biological methods, have been widely applied for characterization, qualitative analysis, and quantitative analysis of dental materials [15]. Biological tests include *in vitro*/systemic toxicity, genotoxicity, carcinogenicity, mutagenicity, reproductive toxicity, immunotoxicity, skin irritation, intracutaneous reactivity, and delayed skin sensitization. The chemical analysis comprises chemical characterization, degradation studies, quantification of metals, and leachable substances [2]. Stino [16] described the application of thermal analysis techniques, namely differential thermal analysis (DTA), differential scanning calorimetry (DSC), and thermogravimetric analysis (TGA) for characterizing dental materials. The principles of premarketing evaluation of the safety and efficacy of dental materials for tooth bleaching in Japan have been reviewed [17].

5.2.1 Quality Control of Dental Materials Using Molecular Spectroscopy

Due to its simplicity, molecular spectroscopy, including ultraviolet-visible spectroscopy, FT-IR spectroscopy, and Raman spectroscopy, has been widely reported for the characterization and analysis of dental materials and for the diagnosis of dental diseases either *in vitro* or *in vivo* [18, 19].

Many review articles regarding the application of spectroscopy methods for the analysis and quality control of dental materials have been published. Infrared (IR) spectroscopy, Fourier transform infrared spectroscopy (FT-IR), Raman spectroscopy, ultraviolet and visible spectroscopy (UV-Vis), X-ray spectroscopy, and mass spectrometry (MS), which were widely employed in the testing of the surface properties of dental materials, have been described and discussed by Kaczmarek et al. [20].

Microstructure, especially fluorescent and chromatic phenomena, appears to play an important role in the ideal teeth and restorative material of dentin. It is now acknowledged that there is a fluorescent phenomenon of the day; ultraviolet radiation makes teeth appear whiter and shinier. The concept of a fluorescent detector is the state of excitation of the atoms of the tooth structure, which, when they return to a state of less excitation, emit light in the visible spectrum between 400 and 450 nm, a range characteristic of blue light. The quantity of UV light falling on the tooth surface is promotional for the emission of fluorescence. Therefore, dental fluorescence becomes more evident at sea level, in the mountains, or in rooms with artificial UV light or black light [21].

The application of UV-visible spectroscopy, fluorescent spectroscopy, IR spectroscopy, FT-IR spectroscopy, near-IR spectroscopy, and Raman spectroscopy in dentistry that are published in 2015–2021 has been reviewed by Indrayanto and Rohman [22]. Lach et al. [23] have published a review article on the application of various spectroscopy methods in implant studies, namely Auger electron spectroscopy (AES), X-ray photoelectron (XPS) spectroscopy, photoluminescence piezo spectroscopy (PLPS), diffuse reflectance FT-IR spectroscopy, fluorescent spectroscopy, X-ray fluorescent spectroscopy, and laser-induced fluorescence [24]. Review articles on the application of Raman spectroscopy techniques for the characterization of dental materials and the diagnostic tools for oral diseases have been published by Zhang et al. [25] and Otel [26]. Kallianta et al. [27] discussed the application of MS in dentistry; MS can be applied for the analysis of biospecimens, as a real-time diagnostic tool in service of oral surgery and pathology, and in research of dental materials and chemical agents. Barrera-Ortega et al. [28] discussed the application of Raman spectroscopy for dental enamel surface characterization. A review article on the application of MS-based proteomic applications in dental implant research has been published by Halstenbach et al. [29]. The application of NMR spectroscopy for medical and dental applications has been reviewed and summarized by Zia et al. [30].

Recently, Beddoe et al. [51] described the application of FT-IR imaging spectroscopy and quantum cascade laser-based infrared (QCL-IR) microscopy for probing the micro- and nanoscopic properties of dental materials. Abbassy et al. [52] reported the characterization of a novel enamel sealer for bioactive remineralization of white spot lesions using ATR-FT-IR, X-ray diffraction (XRD), and scanning electron microscope equipped with electron-dispersive-spectroscopy (SEM/EDS). The measurement of fluoride reaction by bovine enamel and dentin treated with silver diamine fluoride using ^{19}F and ^{31}P solid-state magic angle spinning nuclear magnetic resonance (MAS NMR) has been reported by Hiraishi et al. [31]. Abdulrazzq and Hamad [32] reported the application of FT-IR for the characterization of zein–$CaSiO_3$ composite coating on polyetheretherketone implant material. Seredin et al. [33] described the application of FT-IR spectroscopy and synchrotron FT-IR microspectroscopy to characterize biomimetic composite dental materials composed of nanocrystalline hydroxyapatite and light-cured adhesive; the experiments were performed with a Bruker VERTEX 80v spectrometer coupled with a Hyperion3000 FT-IR microscopy and liquid nitrogen-cooled narrow-band mercury cadmium telluride detector. El-Deeb et al. [34] applied FT-IR and H-NMR spectroscopic methods to characterize synthetic dental adhesives containing synthesized quaternary ammonium compounds. The characterizations of composite resins with different polyhedral oligomeric silsesquioxane additions were performed by using FT-IR spectroscopy as reported by Wang et al. [35]. Iordache et al. [36] described the characterization of three similar dental restorative composite materials using Raman spectroscopy supported by complementary methods such as hardness and density measurements.

These review articles and original articles, which have been cited in this section, showed that various spectroscopic methods can be applied to analyse dental materials for characterization and quality control; unfortunately, the method validations were not described and discussed in detail. Without the validation methods, the integrity of the reported data of the published articles cannot be achieved, so the research or experiment conducted by a particular group of researchers may not yield the same results if repeated by other researchers. Validity ensures that research findings accurately represent the phenomenon studied and can be confidently relied upon. It is recommended that all reported chemical and biological methods should be completed with the validation data.

5.2.2 Ultraviolet-Visible (UV-VIS) Spectroscopy

Some studies have been conducted for quality control of dentals using UV-VIS spectroscopy. Palani and Raguganesh have studied the efficiency of ultra-phosphate or inorganic phosphorus compared with two whitening agents, namely hydrogen peroxide and McInnes solution (a solution combining 30% hydrogen peroxide, 36% hydrochloric acid, and ethyl ether). In this study, sixty human maxillary incisors were selected and decoronated. After that, the samples were subjected to an artificial staining procedure for three days using freshly prepared coffee extract, and UV-VIS spectroscopy was used to evaluate the absorbance changes of samples, and coffee was recorded. The results showed that UV-VIS spectra can be effective tools for the differentiation of samples capable of removing the coffee as observed from the spectral difference between control and samples after being stained with coffee. In this case, ultra-phosphate has a tendency to have higher stain reduction properties (caffeine) when compared to H_2O_2 and McInnes' solution. Ultra-phosphate can bind to the tooth surface because it has a high negative charge and can remove stains from teeth; in addition, the binding energy between ultra-phosphate and teeth is higher than the binding energy between stains and teeth [37]. UV-VIS spectra were also simple tools for monitoring the reaction kinetics of amylase in human saliva needed during forensic investigations of violent crimes [38].

Dental caries is one of the non-communicable diseases having high prevalence trend in recent years. UV-VIS spectroscopy combined with chemometrics has been developed and validated for classification of caries based on its severity. Some spectral preprocessing methods, including mean centring, autoscaling, and Savitzky-Golay smoothing, have been employed to enhance the signal-to-noise ratio of UV-VIS spectral data. In addition, the chemometrics of classification algorithms, including K-nearest neighbours (KNNs), logistic regression (LR), and linear discriminant analysis (LDA), were tried for making the classification of dental caries according to its severity, based on International Caries Detection and Assessment System (ICDAS) scores. An ICDAS score of 0 indicates healthy teeth, score 1 indicates slight discoloration of the enamel, score 2 indicates visual changes to the tooth, score 3 indicates local damage to the enamel, score 4 indicates a dark shadow underlying the dentin, and score 5 indicates a different cavity [39]. Among these, the chemometrics of LDA coupled with Savitzky-Golay smoothing provided the best classification results according to ICDAS scores with respect to the characteristics performance of validation, including accuracy, precision, sensitivity, and specificity. The area under the curve of the ROC plot computed for the LDA algorithm was 0.95, indicating that the prediction model was capable of differentiating normal and carious teeth correctly [40]. Recently, Basri et al. [41] also successfully identified dental caries using UV-VIS spectroscopy in combination with chemometrics of preprocessing spectra and neural networks. Either artificial neural network (ANN) with hyperparameter tuning or convolutional neural network (CNN) combined with Savitzky Golay smoothing provided good validation data in terms of accuracy, precision, sensitivity, and specificity values using variable absorbance values at 270–290 nm. This method allows the classification of dental caries based on ICDAS classification.

5.2.3 Fluorescence Spectroscopy

Fluorescence spectroscopy applying certain excitation and emission wavelengths of electromagnetic radiation (EMR) has been found to be an effective tool for the analysis and detection of dental-related problems and for the characterization of dental materials, as indicated by several reviews applying fluorescence spectroscopy and related techniques for the analysis of dental materials and early detection of dental-related diseases. The interaction between the dentals with EMR can be described by four phenomena, namely specular transmission of the light flux through the dental materials, the specular reflection on the surface, diffuse reflection on the surface, as well as the absorption and scattering within the tooth. In fluorescence spectroscopy, the quantity of UV light falling on the surface of dental materials is promotional to the emission of fluorescence. Therefore, the fluorescence intensity is proportional to the concentration of analytes (dental materials or properties which are related to dental diseases [42]. Macey et al. have reviewed the application of fluorescence spectroscopy for the identification of early caries involving the enamel surface of the tooth [43]. Lee has reviewed the application of fluorescence properties (specific emission spectra) of human teeth and dental calculus intended for clinical applications [44]. The application of fluorescence spectroscopy for the detection of oral cancers has also been reviewed by Raghushaker et al. [45]. The systematic review on the use of autofluorescence for detection of invisible dental biofilm induced by UV light in adult patients has been completely done by Joseph et al. [46].

Fluorescence spectroscopy has been successfully applied for differentiation between dental tissues and resin through the investigation of intensity difference of fluorescence spectra. Resin showed the excitation wavelengths of 250 to 450 nm, while dental tissues exhibited the excitation wavelengths of 250 and 300 nm. In addition, the emission intensity of dental tissues at an emission wavelength of 485 nm and resin having an emission wavelength of 485 nm showed the different intensity significantly. Thus, fluorescence spectroscopy could be an effective and simple analytical method for the assessment of the fluorescence of composite resins and dental tissues [47]. Table 5.1 listed an example of application of fluorescence spectroscopy in dental analysis.

TABLE 5.1 Application of Fluorescence Spectroscopy in Dental Analysis

Authors	Objective	Techniques	Results
Ravazzi et al. [48]	The assessment of the presence of dental biofilm in patients of orthodontic appliances.	Porphyrin-associated fluorescence spectroscopy (Photogen®)	Photogen® was capable of detecting the dental biofilm in the oral environment of patients with orthodontic treatment. The identification of presence of biofilm on the buccal surfaces is better carried using Photogen® than using fluorescence spectroscopy without porphyrin
Wang et al. [49]	Early detection of dental caries and the classification of caries based International Caries and Detection Assessment System (ICDAS)	FS using the excitation light source of 405 nm LED (380–440 nm) and emission light at (440–780 nm). The tooth specimens were captured using the hyperspectral imaging	FS enables the accurate prediction of caries on the entire tooth surface. In addition, the different caries stages could be classified using FS based on ICDAS classification.
Nogueira et al. [50]	The assessment of tooth whitening.	FS using a light-induced method using violet LED illumination	The spectral changes of FS could be used as effective manner for assessment of tooth whitening after the tooth has been stained using coffee. A decrease of 37.4% of total fluorescence intensity at 450 nm (excitation) and 750 nm (emission) after coffee staining.

5.3 CONCLUSION

Molecular spectroscopy based on the interaction between EMR and objects, including dental materials, has been found to be an effective tool for the analysis and detection of dental-related problems and for the characterization of dental materials.

REFERENCES

[1] ISO, "International Organization for Standard, ISO 1942:2020(en) Dentistry – Vocabulary." www.iso.org, 2020. [Online]. Available: https://www.iso.org/obp/ui/en/#iso:std:iso:1.42:ed-3:v1:en:term:3.1.4.8 (accessed April 9, 2024).

[2] G. Schmalz and F. Reichl, "Regulation of Dental Materials," in *Regulatory Toxicology*, F. Reichl and M. Schwenk, Eds., First Edit.Springer, Berlin, Heidelberg, 2021, pp. 3–26. doi: 10.1007/978-3-642-36206-4_126-1 (accessed April 8, 2024).

[3] European Union, "Regulation (EU) 2017/745 of the European parliament and of the council of 5 April 2017 on medical devices, *Official Journal of the European Union L 117* (2017).," no. 5 April 2017. pp. 1–117, 2017. [Online]. Available: https://eur-lex.europa.eu/legal-content/DE/TXT/?uri=CELEX%3A32017R0745 (accessed April 8, 2024).

[4] W. Christina, "Medical Devices: from Directives to Regulations," *BDJ Student*, vol. 27, no. 3, pp. 1–3, 2020.

[5] FIDE, "Classification of dental medical devices according Regulation on Medical Devices 2017/745 (MDR) Annex VIII." 2022. [*Online*]. Available: https://www.fide-online.org/wp-content/uploads/2023/10/Classification_of_dental_devices_according_MDR_2022-02-01-final.pdf (accessed April 8, 2024).

[6] D. Mohn and M. Zehnder, "The medical device regulation (MDR) from a dental perspective," *Front. Dent. Med.*, vol. 4, no. March, 2023, doi: 10.3389/fdmed.2023.1155820

[7] ISO, "International Organization for Standard, ISO 4049: 2019 (en), Dentistry – Polymer-based filling, restorative and luting materials." 2019. [*Online*]. Available: https://www.iso.org/obp/ui/en/#iso:std:iso:4049:ed-5:v1:en (accessed April 9, 2024)

[8] E. (European Regulation), "European Union, Registration, evaluation, authorization and restriction of chemicals (REACH), Official Journal of the European Union L 396/1," *Official Journal of the European Union*. pp. 4–17, 2006. [Online]. Available: https://eur-lex.europa.eu/legal-content/EN/TXT/PDF/?uri=CELEX:32006R1907 (accessed April 8, 2024).

[9] OECD, "OECD Test Guidelines for Chemicals." 2024. [*Online*]. Available: https://www.oecd.org/chemicalsafety/testing/oecdguidelinesforthetestingofchemicals.htm (accessed April 9, 2024).

[10] G. Schmalz, D. C. Watts, and B. W. Darvell, "Dental materials science: Research, testing and standards," *Dent. Mater.*, vol. 37, no. 3, pp. 379–381, 2021, doi: 10.1016/j.dental.2021.01.027

[11] B. W. Darvell, "Misuse of ISO standards in dental materials research," *Dent. Mater.*, vol. 36, no. 12, pp. 1493–1494, 2020, doi: 10.1016/j.dental.2020.10.014
[12] J. L. Brittan, S. V. Sprague, E. L. MacDonald, R. M. Love, H. F. Jenkinson, and N. X. West, "In vivo model for microbial invasion of tooth root dentinal tubules," *J. Appl. Oral Sci.*, vol. 24, no. 2, pp. 126–135, 2016, doi: 10.1590/1678-775720150448
[13] D. Astudillo-Rubio, A. Delgado-Gaete, C. Bellot-Arcís, J. M. Montiel-Company, A. Pascual-Moscardó, and J. M. Almerich-Silla, "Mechanical properties of provisional dental materials: A systematic review and meta-analysis," *PLoS One*, vol. 13, no. 2, pp. 1–19, 2018, doi: 10.1371/journal.pone.0193162
[14] E. E. Maiorov, L. I. Shalamai, E. Y. Mendosa, V. B. Lampusova, and N. S. Oksas, "Determination of the Mechanical Properties of Contemporary Dental Composite Materials by a Stretching Method," *Biomed. Eng. (NY).*, vol. 56, no. 4, pp. 242–246, 2022, doi: 10.1007/s10527-022-10211-5
[15] M. Tabata *et al.*, "Surface analysis of dental caries using a wireless pH sensor and Raman spectroscopy for chairside diagnosis," *Talanta*, vol. 235, no. June, p. 122718, 2021, doi: 10.1016/j.talanta.2021.122718
[16] S. Sherif Stino, "Thermal analysis of dental materials: A review," *Biomat. J.*, vol. 1, no. 2, pp. 13–23, 2022, [*Online*]. Available: http://www.biomatj.com
[17] M. Kurauchi *et al.*, "A Japanese concept of considerations for evaluation of dental materials for tooth bleaching," *Jpn. Dent. Sci. Rev.*, vol. 59, pp. 23–27, 2023, doi: 10.1016/j.jdsr.2023.01.001
[18] A. Rohman and R. T. C. Tandelilin, "Molecular spectroscopy as an analytical tool for the detection of oral cancer," *Oral Diagnostics Tools Tech. A Phys. Approach*, pp. 1–14, 2023, doi: 10.1088/978-0-7503-5316-8ch1
[19] S. Singh, A. Sharma, P. B. Sood, A. Sood, I. Zaidi, and A. Sinha, "Saliva as a prediction tool for dental caries: An in vivo study," *J. Oral Biol. Craniofacial Res.*, vol. 5, no. 2, pp. 59–64, 2015, doi: 10.1016/j.jobcr.2015.05.001
[20] K. Kaczmarek, A. Leniart, B. Lapinska, S. Skrzypek, and M. Lukomska-Szymanska, "Selected spectroscopic techniques for surface analysis of dental materials: A narrative review," *Materials (Basel).*, vol. 14, no. 10, 2021, doi: 10.3390/ma14102624
[21] T. M. Da Silva, H. P. M. De Oliveira, D. Severino, I. Balducci, M. F. R. L. Huhtala, and S. E. P. Gonçalves, "Direct spectrometry: A new alternative for measuring the fluorescence of composite resins and dental tissues," *Oper. Dent.*, vol. 39, no. 4, pp. 407–415, Jul. 2014, doi: 10.2341/12-464-L
[22] G. Indrayanto and A. Rohman, "The Application of Molecular Spectroscopy and Chemometrics in Dentistry," in *Spectroscopic Techniques for Dentistry Application, Recent Advances*, A. K. Shukla, Ed., First Edit.IOP Publishing, Bristol, 2022, pp. 1–30.
[23] S. Lach, P. Jurczak, N. Karska, A. Kubiś, A. Szymańska, and S. Rodziewicz-Motowidło, "Spectroscopic methods used in implant material studies," *Molecules*, vol. 25, no. 3, 2020, doi: 10.3390/molecules25030579
[24] A. Mahmoud, S. Elbasuney, and Y. H. El-Sharkawy, "Instant identification of dental white spot using K-means algorithm via laser-induced fluorescence

and associated hyperspectral imaging," *J. Photochem. Photobiol. B Biol.*, vol. 245, no. June, p. 112749, 2023, doi: 10.1016/j.jphotobiol.2023.112749
[25] Y. Zhang, L. Ren, Q. Wang, Z. Wen, C. Liu, and Y. Ding, "Raman spectroscopy: A potential diagnostic tool for oral diseases," *Front. Cell. Infect. Microbiol.*, vol. 12, no. February, p. 775236, 2022, doi: 10.3389/fcimb.2022.775236
[26] I. Otel, "Overall review on recent applications of raman spectroscopy technique in dentistry," *Quantum Beam Sci.*, vol. 7, no. 1, 2023, doi: 10.3390/qubs7010005
[27] M. Kallianta, E. Pappa, H. Vastardis, and C. Rahiotis, "Applications of mass spectrometry in dentistry," *Biomedicines*, vol. 11, no. 2, 2023, doi: 10.3390/biomedicines11020286
[28] C.C. Barrera-Ortega, A.R.V. Olmos, I.S. Berrú, and D.K. Itzel, "Application of Raman Spectroscopy for Dental Enamel Surface Characterization," in: M. El-Azazy, (Ed.), *Infrared Spectroscopy - Perspectives and Applications*, IntechOpen, 2022, pp. 1–15.
[29] T. Halstenbach, A. Topitsch, O. Schilling, G. Iglhaut, K. Nelson, and T. Fretwurst, "Mass spectrometry-based proteomic applications in dental implants research," *Proteomics - Clinical Applications*, vol. 18, no. 3. pp. 1–16, 2024. doi: 10.1002/prca.202300019
[30] K. Zia, T. Siddiqui, S. Ali, I. Farooq, M. S. Zafar, and Z. Khurshid, "Nuclear magnetic resonance spectroscopy for medical and dental applications: A comprehensive review," *Eur. J. Dent.*, vol. 13, no. 1, pp. 124–128, 2019, doi: 10.1055/s-0039-1688654
[31] M. A. Abbassy, A. S. Bakry, E. H. Almoabady, S. M. Almusally, and A. H. Hassan, "Characterization of a novel enamel sealer for bioactive remineralization of white spot lesions," *J. Dent.*, vol. 109, no. April, p. 103663, 2021, doi: 10.1016/j.jdent.2021.103663
[32] S. Abdulrazzaq and T. Ismael Hamad, "Studying and characterization of coating of Zein-CaSiO3 composite on polyetheretherketone implant material," *Int. J. Dent.*, vol. 2024, 2024, doi: 10.1155/2024/1592439
[33] P. Seredin, D. Goloshchapov, V. Kashkarov, Y. Ippolitov, and J. Vongsvivut, "The molecular and mechanical characteristics of biomimetic composite dental materials composed of nanocrystalline hydroxyapatite and light-cured adhesive," *Biomimetics*, vol. 7, no. 2, 2022, doi: 10.3390/biomimetics7020035
[34] M. El-Deeb, M. M. Ismail, E. R. Kenawy, and N. A. Habib, "Evaluation of antibacterial activity of an experimental dental adhesive containing synthesized quaternary ammonium compound: in vitro study," *Discov. Appl. Sci.*, vol. 6, no. 3, 2024, doi: 10.1007/s42452-024-05756-x
[35] J. Wang, Y. Liu, J. Yu, Y. Sun, and W. Xie, "Study of POSS on the properties of novel inorganic dental composite resin," *Polymers (Basel).*, vol. 12, no. 2, pp. 1–10, 2020, doi: 10.3390/polym12020478
[36] S. M. Iordache *et al.*, "Performance assessment of three similar dental restorative composite materials via raman spectroscopy supported by complementary methods such as hardness and density measurements," *Polymers (Basel).*, vol. 16, no. 4, 2024, doi: 10.3390/polym16040466

[37] S. K. Palani and V. Raguganesh, "Ultraviolet–visible spectrophotometric analysis of tooth whitening efficiency of ultraphosphate," *J. Contemp. Med. Dent.*, vol. 8, no. 1, p. 48, 2020, [Online]. Available: www.jcmad.com
[38] T. L. Reena Roy, "Ultraviolet-Visible Spectrophotometry (UV-VIS) and SALIgAE® qualitative and semi-quantitative tools for the analysis of salivary amylase," *J. Forensic Res.*, vol. 05, no. 06, 2014, doi: 10.4172/2157-7145.1000247
[39] A. I. Ismail *et al.*, "The International Caries Detection and Assessment System (ICDAS): An integrated system for measuring dental caries," *Comunity Dent. oral Epidemiol.*, vol. 35, no. 3, pp. 170–178, 2007, doi: 10.1111/j.1600-0528.2007.00347x
[40] K. N. Basri, F. Yazid, R. Megat Abdul Wahab, M. N. Mohd Zain, Z. Md Yusof, and A. S. Zoolfakar, "Chemometrics analysis for the detection of dental caries via UV absorption spectroscopy," *Spectrochim. Acta - Part A Mol. Biomol. Spectrosc.*, vol. 266, p. 120464, 2022, doi: 10.1016/j.saa.2021.120464
[41] K. N. Basri, F. Yazid, M. N. Mohd Zain, Z. Md Yusof, R. Abdul Rani, and A. S. Zoolfakar, "Artificial neural network and convolutional neural network for prediction of dental caries," *Spectrochim. Acta - Part A Mol. Biomol. Spectrosc.*, vol. 312, no. January, p. 124063, 2024, doi: 10.1016/j.saa.2024.124063
[42] A. Ribeiro *et al.*, "A preliminary investigation of a spectroscopic technique for the diagnosis of natural caries lesions," *J. Dent.*, vol. 33, no. 1, pp. 73–78, 2005, doi: 10.1016/j.jdent.2004.08.006
[43] R. Macey *et al.*, "Fluorescence devices for the detection of dental caries," *Cochrane Database Syst. Rev.*, vol. 2020, no. 12, 2020, doi: 10.1002/14651858.CD013811
[44] Y.-K. Lee, "Fluorescence properties of human teeth and dental calculus for clinical applications," *J. Biomed. Opt.*, vol. 20, no. 4, p. 040901, 2015, doi: 10.1117/1.jbo.20.4.040901
[45] C. R. Raghushaker, M. D'Souza, A. S. Urala, S. Ray, and K. K. Mahato, "An overview of conventional and fluorescence spectroscopy tools in oral cancer diagnosis," *Lasers Dent. Sci.*, vol. 4, no. 4, pp. 167–179, 2020, doi: 10.1007/s41547-020-00104-9
[46] B. Joseph *et al.*, "Detection of invisible dental biofilm using light-induced autofluorescence in adult patients–A systematic review," *Photodiagnosis Photodyn. Ther.*, vol. 39, no. April, p. 102916, 2022, doi: 10.1016/j.pdpdt.2022.102916
[47] T. M. Da Silva, H. P. M. De Oliveira, D. Severino, I. Balducci, M. F. R. L. Huhtala, and S. E. P. Gonçalves, "Direct spectrometry: A new alternative for measuring the fluorescence of composite resins and dental tissues," *Oper. Dent.*, vol. 39, no. 4, pp. 407–415, 2014, doi: 10.2341/12-464-L
[48] R. Ravazzi, J. G. Neves, M. P. Santamaria, L. P. Rosa, F. C. S. Rosa, and M. Santamaria-Jr, "Porphyrin-associated fluorescence spectroscopy (Photogen®) for the optical diagnosis of dental biofilm in orthodontic treatment: An observational clinical trial," *Photodiagnosis Photodyn. Ther.*, vol. 42, no. February, 2023, doi: 10.1016/j.pdpdt.2023.103580

[49] C. Wang *et al.*, "Fluorescence spectrometry based chromaticity mapping, characterization, and quantitative assessment of dental caries," *Photodiagnosis Photodyn. Ther.*, vol. 37, no. October 2021, p. 102711, 2022, doi: 10.1016/j.pdpdt.2021.102711

[50] M. Saito Nogueira, V. H. Panhóca, and V. S. Bagnato, "Fluorescence spectroscopy analysis of light-induced tooth whitening," *Proc. SPIE*, vol. 10876, p. 45, 2019, doi: 10.1117/12.2507665

[51] M. Beddoe, T. Gölz, M. Barkey, E. Bau, M. Godejohann, S.A. Maier, F. Keilmann, M. Moldovan, D. Prodan, N. Ilie, and A. Tittl, Probing the micro- and nanoscopic properties of dental materials using infrared spectroscopy: A proof-of-principle study, *Acta Biomater* vol. 168, pp. 309–322, 2023. https://doi.org/10.1016/j.actbio.2023.07.017

[52] N. Hiraishi, M. Sayed, R. Hill, and Y. Shimada, Solid-state NMR spectroscopy measurement of fluoride reaction by bovine enamel and dentin treated with silver diammine fluoride, *Dental Mater.* vol. 38, pp. 769–777, 2022. https://doi.org/10.1016/j.dental.2022.04.017

CHAPTER 6

Mass Spectrometry for Analysis of Dental Materials

Nuray Bağcı and Zühre Akarslan

6.1 INTRODUCTION

Mass spectrometry is a commonly used spectroscopic technique for compound analysis in laboratories. It was introduced to the field in 1912 with the construction of the mass spectrometer (1). This technique can determine a molecule's chemical and structural properties, identify unknown compounds, quantify known compounds, and measure isotopes and molecular masses. Gas or liquid chromatographs are added to this method to achieve more accurate and detailed results during analysis.

Clinical behavior, toxicity, and antimicrobial properties of dental materials, such as composites, polymers, biomaterials, metals, and ceramics, could be analyzed with mass spectrometry. It is a very sensitive method and can even detect tiny amounts (femtograms) of the compounds present in a sample (2).

In mass spectrometry, the analysis starts with the ionization of the molecules/components of the sample. Mainly, somebody could make the sample ionization with the following:

1. Secondary ion mass spectrometry
2. Electron impact ionization
3. Fast atom bombardment

DOI: 10.1201/9781003506805-6

4. Electrospray ionization
5. Laser desorption ionization
6. Matrix-assisted laser desorption ionization
7. Inductively coupled plasma

(2–5)

During the ionization process, the molecules broke into smaller ions. Then, these ions start to rush according to their mass and charge ratio (m/z). A specific detector captures and translates them into an electric signal. In the end, a plot of intensity as a function of the mass-to-charge ratio gives the data. This process allows the identification of the chemical compounds in the material investigated (2,6).

In this technique, the operator should prepare the sample carefully according to the directions and use appropriate measuring tools for an accurate analysis. The mass spectrometer can reach high temperatures, and ionization is harmful; thus, the operator should wear protective equipment while using the device (7).

Secondary ion mass spectrometry: This technique is among the most used techniques during dental material analysis. The operator could analyze a material's outer surface and sub-surface composition with secondary ion mass spectrometry.

In this technique, a high-energy primary beam is focused on the material's outer surface. This beam sputters the specific atoms and molecules of the material under vacuum. Most removed atoms and molecules from the surface are neutral, and some are ionized. The injected beam of secondary ions from this procedure is accelerated, focused, and analyzed with a mass spectrometer. Finally, we obtain a mass spectrum, depth profile, and sample image (8–10). The operator should be careful about the composition and surface features of the material. The material going to be analyzed must be resistant to high vacuum, should not contain any loose parts, and should be flat (11). The main limitation of this technique is that the high-energy primary beams quickly sputter away from the sample's surface.

Time of flight secondary ion mass spectrometry: Qualitative surface analysis of the elements and molecules is made with this technique. It is a proper confirmatory method and can detect the onset of a chemical reaction.

This method uses a drift tube to detect the sputtered atoms. The time of flight of a fragment is calculated, and this is used for the determination of the mass. This method makes the technique sensitive and allows depth profiling and mapping; however, it damages the sample (12).

Laser ablation inductively coupled plasma mass spectrometry (LA-ICP-MS): This is a kind of ion spectroscopy useful for assessing the mass of a sample. The operator could investigate elements in solid samples with this method. The time of flight of the charged fragments in the material gives the sample properties. Depth profile and mapping of the specimen can be made. However, the sample is destroyed during the analysis.

In this method, the sample is vaporized with an Nd:YAG laser source, and the fragments are transported into a high-energy plasma of an ICP mass spectrometer. Subsequently, the composition of the material is mapped (12,13).

6.2 MASS SPECTROMETRY FOR DENTAL MATERIALS

The operator could investigate the characteristics of dental materials with mass spectrometry. The structure and leachability of resin composite and amalgam, the surface of zirconia, ceramic, and implants, and components of bioceramic are the mainly analyzed characteristics of these materials. Thus, the clinical behavior of dental materials and their impact on human health could be evaluated (2,11).

Residual monomer and additive components can elute after dental composite resin polymerizes. Therefore, many studies have focused on the leachable residual monomer and additive components. In addition, studies assessed different elements, such as monomers, photo-initiators, and stabilizing agents in dental composites.

Spahl et al. investigated possible leachable residual monomer and additive components from four dental composites after polymerization using gas and liquid chromatography/mass spectrometry. They reported that they identified monomers and various additive components in all polymerized composite resin specimens (14).

Using gas chromatography-mass spectrometry, Manabe et al. aimed to determine concentrations of Bisphenol-A, the potential xenoestrogens of eluted resin in dental materials. They showed that minimal concentrations of Bisphenol-A were released from dental materials (15).

Nilsen et al. analyzed organic components in resin-modified pulp capping materials and safety data sheets using gas chromatography-mass spectrometry and ultra-performance liquid chromatography-mass spectrometry.

They concluded that various organic compounds elute and were non-affirmed in the sheets of these dental materials (16).

Vervliet et al. qualitatively analyzed several dental material ingredients, unpolymerized composite resins, and a frequently used dental sealant using liquid chromatography coupled to quadrupole time of flight mass spectrometry. They detected significant components, diverse isomers, and impurities of the dental material ingredients. Non-stated major components, such as Bisphenol-A ethoxylate methacrylate (BisEMA), Bisphenol-A glycidyl methacrylate (BisGMA), and trimethylene glycol dimethacrylate (TEGDMA) in the material safety data sheets, were identified (17).

Vervliet et al. investigated the in vitro metabolism of the dental resin monomers BisGMA, BisPMA, TCD-DI-HEA, and urethane dimethacrylate (UDMA) using human liver microsomes and quadrupole time of flight mass spectrometry. They identified BisGMA, BisPMA, TCD-DI-HEA, and UDMA biotransformation products. In addition, the possible human health effects of these biotransformation products are still unclear (18).

Simkova et al. examined Bisphenol-A or alternative bisphenols of dental restorative composites using liquid chromatography-tandem mass spectrometry. They found that Bisphenol-A was released from both 'Bisphenol-A-free' composites and 'BisGMA-containing' composites. It has been notified that the amount of Bisphenol-A is very low and does not pose a health risk if the dental composite is adequately polymerized (19).

In the study of Barisic et al., the researchers used liquid chromatography coupled with triple quadrupole tandem mass spectrometry to investigate leachable components (monomers) in high- and low-viscosity dental bulk-fill resin composites and conventional resin composite materials after polymerization. They found that monomers HEMA, TEGDMA, 4-dimethylaminobenzoic acid ethyl ester (DMA BEE), and BisGMA in bulk-fill and conventional composites (high and low viscosity) can be eluted after polymerization (20).

Another study aimed to determine the proteomic profile of salivary pellicles formed on different dental composites. The saliva and pellicle samples were analyzed by SELDI-Time-of-flight secondary ion mass spectrometry. It has been reported that SELDI-Time-of-flight secondary ion mass spectrometry identified up to 102 different proteins in the saliva samples and up to 46 proteins in the pellicle (21).

In addition to composite and resin, amalgam was also analyzed by mass spectrometry. Tercanli et al. investigated the effect of amalgam restoration on mercury concentration in saliva using inductively coupled plasma mass spectrometry. They found that the mercury concentration in unstimulated saliva was higher in the group with amalgam than in the group without amalgam (22).

Surfaces of dental materials such as zirconia and ceramic were examined with mass spectrometry.

Using time-of-flight secondary ion mass spectrometry, Chuang et al. evaluated the effects of silane- and MDP-based primer application orders on zirconia-resin adhesion. MDP-based primer has been shown to facilitate resin-zirconia adhesion (23).

Lima et al. investigated the effect of silane- and MDP-based primers on the physico-chemical properties of zirconia and its bond strength to resin cement using time-of-flight secondary ion mass spectrometry. They found that the MDP-based primers group showed more chemical linkage compared to the other groups (24).

Lapinska et al. researched the effects of surface cleaning procedures on glass ceramic adhesion using time-of-flight secondary ion mass spectrometry. In conclusion, they suggested that the most effective cleaning procedure for saliva-contaminated ceramic surfaces was cleaning leucite glass-ceramic surfaces with orthophosphoric acid or re-etching the lithium disilicate glass-ceramic surfaces with hydrofluoric acid (25).

Möncke et al. developed a new lithium disilicate glass-ceramic spray-coated on ZrO_2-based dental restoration. They used time-of-flight secondary ion mass spectrometry to examine the crystallization and surface behaviors (26).

A few studies investigated the surface of dental implants using mass spectrometry. Saha et al. aimed to identify oral cavity bacterial colonization around two-piece dental implants. The bacterial colonies were identified with matrix-assisted laser desorption ionization-time of flight mass spectrometry. They reported that the method was adequate for the identification of potential pathogenic bacteria around two-piece dental implants (27).

Zuanazzi et al. evaluated the possible effect of different titanium modifications on salivary pellicle content by analyzing the salivary proteome through mass spectrometry-based proteomics. The researchers found that bio-integration was affected by salivary proteins adsorbed on titanium surfaces (28).

A recent study investigated the impurities in commercial titanium dental implants using inductively coupled plasma-mass spectrometry and optical emission spectrometry. It has been revealed that titanium dental

TABLE 6.1 Use of Mass Spectrometry for Dental Materials

Authors	Year	Sample	Mass spectrometry	Outcome
Spahl et al.	1998	Resin composite	Gas and liquid chromatography/mass spectrometry	Monomers and various additive components identified in all polymerized composite resin specimens.
Manabe et al.	2000	Resin composite	Gas chromatography-mass spectrometry	A minimal concentration of Bisphenol-A can be released from dental materials.
Nilsen et al.	2017	Resin-modified pulp capping materials	Gas chromatography-mass spectrometry and ultra-performance liquid chromatography-mass spectrometry	Various organic compounds elute which are not affirmed in the sheets of these dental materials.
Vervliet et al.	2018	Dental material ingredients, unpolymerized composite resins, dental sealant	Liquid chromatography coupled to quadrupole time of flight mass spectrometry	Major components such as Bisphenol-A ethoxylate methacrylate (BisEMA), Bisphenol-A glycidyl methacrylate (BisGMA), and triethylene glycol dimethacrylate (TEGDMA) and various isomers and impurities of the dental material ingredients have been detected.
Vervliet et al.	2019	Dental resin monomers	Quadrupole time of flight mass spectrometry	Various amounts of biotransformation products were identified for BisGMA, BisPMA, TCD-DI-HEA, and UDMA.
Simkova et al.	2020	Resin composite	Liquid chromatography-tandem mass spectrometry	Bisphenol-A was released from both 'Bisphenol-A-free' composites and 'BisGMA-containing' composites.
Barisic et al.	2023	Resin composite	Liquid chromatography coupled with triple quadrupole tandem mass spectrometry	Monomers HEMA, TEGDMA, 4-dimethylaminobenzoic acid ethyl ester (DMABEE) and BisGMA in bulk-fill and conventional composites (high and low viscosity) can be eluted after polymerization.

Reise et al.	2023	Resin composite	SELDI-Time-of-flight secondary ion mass spectrometry	Up to 102 different proteins in the saliva samples and up to 46 proteins in the pellicle were identified.
Tercanli et al.	2024	Amalgam	Inductively coupled plasma mass spectrometry	The mercury concentration in unstimulated saliva was higher in the group with amalgam than in the group without amalgam.
Chuang et al.	2017	Silane- and MDP-based primers- zirconia adhesion surface	Time-of-flight secondary ion mass spectrometry	MDP-based primer has been shown to facilitate resin-zirconia adhesion.
Lima et al.	2019	Silane- and MDP-based primers- zirconia adhesion surface	Time-of-flight secondary ion mass spectrometry	MDP-based primers group showed more chemical linkage than the other groups.
Lapinska et al.	2019	Ceramic	Time-of-flight secondary ion mass spectrometry	The most effective cleaning procedure for saliva-contaminated ceramic surface has been suggested.
Möncke et al.	2017	ZrO_2-based dental restoration	Time-of-flight secondary ion mass spectrometry	The crystallization and surface behaviors were examined.
Saha et al.	2020	Dental implants	Matrix-assisted laser desorption ionization-time of flight mass spectrometry	Mass spectrometry can be effective for the identification of potential pathogenic bacteria around two-piece dental implants.
Zuanazzi et al.	2021	Different titanium surface modifications	Mass spectrometry-based proteomics	Biointegration is affected by many salivary proteins adsorbed on titanium surfaces.
Stricker et al.	2022	Dental implants	Inductively coupled plasma-mass spectrometry and optical emission spectrometry	Titanium dental implants are potentially allergenic and contain toxic metals and radionuclides.
França et al.	2014	Bioceramics	Time-of-flight secondary ion mass spectroscopy	The differences between the substances investigated were at the molecular level.
Özkan and Gök	2023	Clear aligner orthodontic materials	Liquid chromatography-mass spectrometry/mass spectrometry	The amount of released Bisphenol-A, Bisphenol-F, and Bisphenol-S was considered harmless.

implants were potentially allergenic and contained noxious metals and radionuclides (29).

Studies analyzed bioceramics and orthodontic materials with mass spectrometry. Using time-of-flight secondary ion mass spectroscopy, França et al. investigated the mechanisms of calcium phosphates, such as bioactive calcium hydroxyapatite, biphasic calcium, and b-tricalcium phosphates. The differences of the substances investigated were at the molecular level (30). Özkan and Gök evaluated bisphenol release from different clear aligner materials with liquid chromatography-mass spectrometry/mass spectrometry. They reported that the amount of released Bisphenol-A, Bisphenol-F, and Bisphenol-S was at a level considered harmless to health (31).

Considering these studies, mass spectrometry is widely used for the analysis of dental materials. A summary of these studies is presented in Table 6.1.

REFERENCES

1. El-Aneed, A., Cohen, A. M., Banoub, J. Mass Spectrometry, Review of the Basics: Electrospray, MALDI, and Commonly Used Mass Analyzers. *Applied Spectroscopy Reviews* 2009, 44(3), 210–230. 10.1080/05704920902717872
2. Kallianta, M.; Pappa, E.; Vastardis, H.; Rahiotis, C. Applications of Mass Spectrometry in Dentistry. *Biomedicines* 2023, 11, 286. 10.3390/biomedicines11020286
3. Yahia, L.H.; Mireles, L.K. X-ray photoelectron spectroscopy (XPS) and time-of-flight secondary ion mass spectrometry (ToF SIMS). In *Characterization of Polymeric Biomaterials*; Elsevier: New York, NY, USA, 2017; pp. 83–97. ISBN 9780081007372.
4. Gosetti, F.; Marengo, E. Mass spectrometry. Selected Ion Monitoring. In *Encyclopedia of Analytical Science*; Elsevier: New York, NY, USA, 2019; pp. 500–510. ISBN 9780081019832
5. Schaepe, K.; Jungnickel, H.; Heinrich, T.; Tentschert, J.; Luch, A.; Unger, W.E.S. Secondary Ion Mass Spectrometry. In *Characterization of Nanoparticles*; Elsevier: New York, NY, USA, 2020; pp. 481–509. ISBN 9780128141830.
6. Aitekenov S.; Gaipov A. Bukasov R Review: Detection and Quantification of Proteins in Human Urine. *Talanta*. 2021 Feb 1, 223(Pt 1), 121718. doi: 10.1016/j.talanta.2020.121718
7. Garg E.; Zubair M. Mass Spectrometer. NCBI Bookshelf. A Service of the National Library of Medicine, *National Institutes of Health. StatPearls [Internet]*. Treasure Island (FL): StatPearls Publishing; 2024.
8. Fitzsimons I.C.W.; Harte B.; Clark R.M. SIMS Stable Isotope Measurement: Counting Statistics and Analytical Precision. *Mineralogical Magazine* 2000, 64, 59–83

9. Matthiesen R.; Bunkenborg J. Introduction to Mass Spectrometry-based Proteomics. *Methods in Molecular Biology* 2013, 1007, 1–45.
10. Alexandrov T.; Chernyavsky I.; Becker M.; von Eggeling F.; Nikolenko S. Analysis and Interpretation of Imaging Mass Spectrometry Data by Clustering Mass-to-Charge Images According to their Spatial Similarity. *Analytical Chemistry* 2013 Dec 03, 85(23), 11189–11195.
11. Kaczmarek, K.; Leniart, A.; Lapinska, B.; Skrzypek, S.; Lukomska-Szymanska, M. Selected Spectroscopic Techniques for Surface Analysis of Dental Materials: A Narrative Review. *Materials* 2021, 14, 2624. https://doi.org/10.3390/ma14102624
12. Sacher E.; Franca, R., *Surface Analysis Techniques for Dental Materials*, World Scientific. Singapore pp. 1–31, 2018.
13. Koch, J.; Günther, D. Laser Ablation Inductively Coupled Plasma Mass Spectrometry. In: *Encyclopedia of Spectroscopy and Spectrometry*. Editors. Lindon JC, Tranter GE, Koppenaal DW. Elsevier 2017, pp. 526–532 (Third Edition)
14. Spahl, W.; Budzikiewicz, H.; Geurtsen, W. Determination of Leachable Components from Four Commercial Dental Composites by Gas and Liquid Chromatography/Mass Spectrometry. *Journal of Dentistry*, 1998, 26, 137–145.
15. Manabe, A.; Kaneko, S.; Numazawa, S.; Itoh, K.; Inoue, M.; Hisamitsu, H.; Sasa, R.; Yoshida, T. Detection of Bisphenol-A in Dental Materials by Gas Chromatography-Mass Spectrometry. *Dental Materials Journal*, 2000, 19, 75–86.
16. Nilsen, B.W.; Jensen, E.; Örtengren, U.; Michelsen, V.B. Analysis of Organic Components in Resin-Modified Pulp Capping Materials: Critical Considerations. *European Journal of Oral Science*, 2017, 125, 183–194.
17. Vervliet, P.; De Nys, S.; Boonen, I.; Duca, R. C.; Elskens, M.; Van Landuyt, K. L.; & Covaci, A. Qualitative Analysis of Dental Material Ingredients, Composite Resins and Sealants Using Liquid Chromatography Coupled to Quadrupole Time of Flight Mass Spectrometry. *Journal of Chromatography A*, 2018, 1576, 90–100.
18. Vervliet, P.; Van Den Plas, J.; De Nys, S.; Duca, R. C.; Boonen, I.; Elskens, M.; Landuyt K. l. v.; Covaci, A. Investigating the In Vitro Metabolism of the Dental Resin Monomers BisGMA, BisPMA, TCD-DI-HEA and UDMA Using Human Liver Microsomes and Quadrupole Time of Flight Mass Spectrometry. *Toxicology*, 2019, 420, 1–10.
19. Šimková, M.; Tichý, A.; Dušková, M.; Bradna, P. Dental Composites–a Low-dose Source of Bisphenol A?. *Physiological Research*, 2020, 69, 295–304.
20. Barišić, M. L.; Sarajlija, H ; Klarić, E.; Knežević, A.; Sabol, I.; Pandurić, V. Detection of Leachable Components from Conventional and Dental Bulk-Fill Resin Composites (High and Low Viscosity) Using Liquid Chromatography-Tandem Mass Spectrometry (LC-MS/MS) Method. *Polymers*, 2023, 15, 627–643.
21. Reise, M.; Kranz, S.; Heyder, M.; Beck, J.; Roth, C.; Guellmar, A.; Eggeling, F.; Schubert, U.; Löffler, B.; Sigusch, B. Salivary Pellicle Formed on Dental

Composites Evaluated by Mass Spectrometry—An In Situ Study. *Molecules*, 2023, 28, 6804–6821.

22. Tercanli, H.; Yavuz, E.; Yilmaz, S. G.; Yardimci, S. Mercury Concentration in Saliva and the Impact of Chewing: An Inductively Coupled Plasma Mass Spectrometry Study. *Biological Trace Element Research*, 2024, 202, 1965–1971.
23. Chuang, S. F.; Kang, L.L.; Liu, Y. C.; Lin, J. C.; Wang, C. C.; Chen, H. M.; Tai, C. K. Effects of Silane-and MDP-Based Primers Application Orders on Zirconia–Resin Adhesion—A ToF-SIMS Study. *Dental Materials*, 2017, 33, 923–933.
24. Lima, R.B.W.; Barreto, S.C.; Alfrisany, N.M.; Porto, T.S.; De Souza, G.M.; De Goes, M.F. Effect of Silane and MDP-Based Primers on Physico-Chemical Properties of Zirconia and Its Bond Strength to Resin Cement. *Dental Materials*, 2019, 35, 1557–1567.
25. Lapinska, B.; Rogowski, J.; Nowak, J.; Nissan, J.; Sokolowski, J.; Lukomska-Szymanska, M. Effect of Surface Cleaning Regimen on Glass Ceramic Bond Strength. *Molecules*, 2019, 24, 389–406.
26. Möncke, D.; Ehrt, R.; Palles, D.; Efthimiopoulos, I.; Kamitsos, E.I.; Johannes, M. A Multi Technique Study of a New Lithium Disilicate Glass-ceramic Spray-coated on ZrO2 Substrate for Dental Restoration. *Biomedical Glases*, 2017, 3, 41–55.
27. Saha, S.; Meenawat, A.; Sahu, C.; Srivastava, V.; Yadav, S.; Kumar, V. Bacterial Identification and Monitoring Around Two-Piece Dental Implants by Matrix-Assisted Laser Desorption Ionization Time of Flight Mass Spectrometry (MALDI-TOF MS). *Journal of Laboratory Physicians*, 2020, 12, 49–55.
28. Zuanazzi, D.; Xiao, Y.; Siqueira, W.L. Evaluating Protein Binding Specificity of Titanium Surfaces through Mass Spectrometry- Based Proteomics. *Clinical Oral Investigation*, 2021, 25, 2281–2296.
29. Stricker, A.; Bergfeldt, T.; Fretwurst, T.; Addison, O.; Schmelzeisen, R.; Rothweiler, R.; Nelson, K.; Gross, C. Impurities in Commercial Titanium Dental Implants - A Mass and Optical Emission Spectrometry Elemental Analysis. *Dental Materials*, 2022, 38, 1395–1403.
30. França, R.; Samani, T.D.; Bayade, G.; Yahia, L.; Sacher, E. Nanoscale Surface Characterization of Biphasic Calcium Phosphate, with Comparisons to Calcium Hydroxyapatite and β-tricalcium Phosphate Bioceramics. *Journal of Colloid Interface Science*, 2014, 420, 182–188.
31. Çifçi Özkan, E.; Dumanlı Gök, G. Evaluation of Bisphenol Release of Different Clear Aligner Materials Using the Liquid Chromatography-mass Spectrometry/mass Spectrometry Method. *The Angle Orthodontist*, 2023, 93(6), 721–726.

CHAPTER 7

Advanced Mass Spectrometry for Dental Materials

Integrated with Chromatography

Halida Rahmania, Eka Noviana,
Agustina Ari Murti Budi Hastuti,
Siti Nurul Hidayah, and Abdul Rohman

7.1 INTRODUCTION

7.1.1 Overview of Mass Spectrometry Principles and History of Dental Material Analysis Using MS

Mass spectrometry (MS) has rapidly evolved as a vital technique in analytical instrumentation. It is a robust method used to identify and quantify compounds by measuring their mass-to-charge ratios (*m/z*). The core principles of MS involve ionizing chemical species, separating these ions based on their *m/z* ratios, and detecting them with high precision [1].

MS has long been recognized as a dependable method for material evaluation. Its origins trace back to Sir J.J. Thomson's work centuries ago, where he determined the *m/z* value of gaseous ionized molecules, providing the first evidence of isotopes [2–4]. Since then, MS technology has significantly advanced. Early ionization methods, such as electric discharge and electron impact (EI), enabled researchers to evaluate the *m/z* of

DOI: 10.1201/9781003506805-7

substances. Initially, the focus was on analyzing sugars, alkaloids, and peptides [5].

The application of MS in dental research gained momentum in the mid-20th century. Early studies focused on analyzing the chemical composition of teeth and dental tissues [6,7]. Researchers evaluated the mineral content of enamel and dentin, advancing the understanding of dental materials' impact on health and disease. Additionally, gas chromatography-mass spectrometry (GC-MS) was utilized to study organic compounds in dental materials, aiding in the identification of potential contaminants in dental cements and resins [8].

The integration of MS with liquid chromatography (LC-MS) marked a significant advancement, offering better separation and higher sensitivity for a wider range of compounds [9]. Researchers used this method to study chemical stability and potential degradation products, providing a breakthrough in evaluating the safety of dental materials [10]. During the 1990s and 2000s, LC-MS became pivotal in dental research. For instance, amalgam, primarily composed of Ag_3Sn (γ) with some Cu and Zn, was used for teeth restoration to minimize dimensional changes during setting [11]. However, its use significantly decreased after 2013 due to concerns about mercury (Hg) release and environmental impact [12,13]. MS studies evaluated the safety of amalgam and paved the way for more stability and safety evaluations of dental materials using LC-MS.

Tandem MS (MS/MS) techniques then gained prominence, allowing for a deeper understanding of molecular structures and trace elements with unparalleled accuracy and precision [14]. MS was employed to investigate fluoride incorporation into enamel and the release of metal ions from dental amalgams and implants [15,16]. MS/MS also enabled the study of interactions between dental materials and biological tissues [17].

The advancements in MS analysis are later continued. High-resolution MS combined with soft ionization techniques like electrospray ionization (ESI) facilitates the efficient characterization of atomically precise clusters [18]. ESI, along with the development of the time-of-flight (TOF) technique, enhanced resolving power ($m/\Delta m$) [19,20]. In a recent study, TOF secondary ion mass spectrometry (TOF-SIMS) is utilized for qualitative chemical surface analysis in dentistry to study surface coatings, ceramic surface contamination, regenerative mineralization of biomedical implants, and zirconium phosphate compounds, covering diverse samples from biological specimens to electronic devices [21]. As MS technology advances, its contributions to dental research are expected to grow, leading to safer, more effective, and innovative dental treatments.

7.1.2 Importance and Relevance of MS in Modern Dental Material Analysis

MS has emerged as a critical tool for evaluating contemporary dental materials. The development of dental materials is a continuous process driven by the evolving demands of the dental field. Historically, the quest to improve dental care has spanned centuries, with ongoing efforts to find optimal materials for restoring and replacing tooth structures [22]. Today, the focus in dentistry has expanded to include aesthetic considerations alongside health-related purposes for restorative procedures [23,24].

Dental materials used in the oral cavity must meet stringent physicochemical and biological requirements due to their direct contact with tissues. These materials need to be biocompatible and non-toxic both locally and systemically [25], while also being able to withstand the physical and chemical challenges of the oral environment [26]. They must maintain biophysical neutrality, accurately replicate natural tooth shapes, offer mechanical stability, be easy to process, provide aesthetic appeal, and remain cost-effective [27,28].

Preventing micro-leakage and biofilm formation is essential, which requires a deep understanding of the material's structural and chemical components [29]. Key attributes such as surface structure, roughness, chemical composition, and reactivity must be thoroughly evaluated to determine the material's suitability for dental application [30]. These surface characteristics are interconnected with other properties, including mechanical and chemical features, which are vital for predicting the material's performance in the oral environment [31].

As a highly sensitive and versatile analytical tool, MS can identify both known and unknown compounds, making it invaluable in dental material analysis. Its applications in this field include assessing the clinical behavior, potential toxicity, and antimicrobial properties of these materials. The insights gained from MS analysis play a significant role in developing safer and more effective dental materials.

7.2 CHROMATOGRAPHY-MASS SPECTROMETRY

Hyphenated techniques combine the advantages of multiple techniques to achieve powerful analytical performance. These techniques are especially useful for analyzing complex samples or materials, including those used in dentistry. Hyphenating chromatography or other separation techniques with spectroscopic/spectrometric methods is common to enable (1) sensitive and selective detection of analyte(s) of interest, (2) simultaneous analysis of multi-components that are present in the samples, and/or (3) determination of chemical compositions [32–34]. While techniques such as tandem mass spectrometry (MS/MS) or high-resolution MS by themselves can provide information on the chemical compositions of the samples, direct MS analysis often suffers from disadvantages such as ionization suppression or enhancement. The presence of many compounds in complex samples at the same time in the MS ion source can cause ionization competition [35,36]. This issue can be minimized by coupling the MS with separation techniques, which allow multiple compounds in the sample to enter the ion source at different times for better ionization and detection.

Chromatography-MS techniques such as GC-MS and LC-MS are widely used for analyzing dental materials [32–34, 37,38]. GC-MS is frequently used for analyzing volatile and thermostable molecules such as monomers and other small, leachable compounds from resin-based dental materials [37–40], while LC-MS can be used for qualitative and quantitative analysis of a wider array of dentistry-relevant compounds/materials, including leachable components from dental resins [32–34], active pharmaceutical ingredients in dental products [41], illicit drugs in dental tissues or oral fluid [42,43], and proteins in dental tissues [44]. More detailed examples of GC-MS and LC-MS applications on these compounds/materials will be discussed in the next sections. In a chromatography-MS system, the sample is typically injected/introduced into the chromatography system, followed by separation based on compounds' polarity (in LC) or the combination of their polarity and boiling points (in GC), ionization of the compounds in the MS source, ion separation based on their mass-to-charge ratios (*m/z*) in the mass analyzer, and finally detection of the ions (Figure 7.1).

In GC-MS, the sample is vaporized and separated into its components using a capillary column coated with a solid/liquid stationary phase. Separated compounds are then ionized via electron impact/electron ionization (EI), chemical ionization (CI), or field ionization (FI). EI is the most common in GC-MS, in which ions are created from the interaction between the gaseous molecules and a high-energy (up to 70 eV) electron beam [38, 39]. This hard ionization produces multiple fragment ions that can aid the identification of the molecules. Both CI and FI are soft

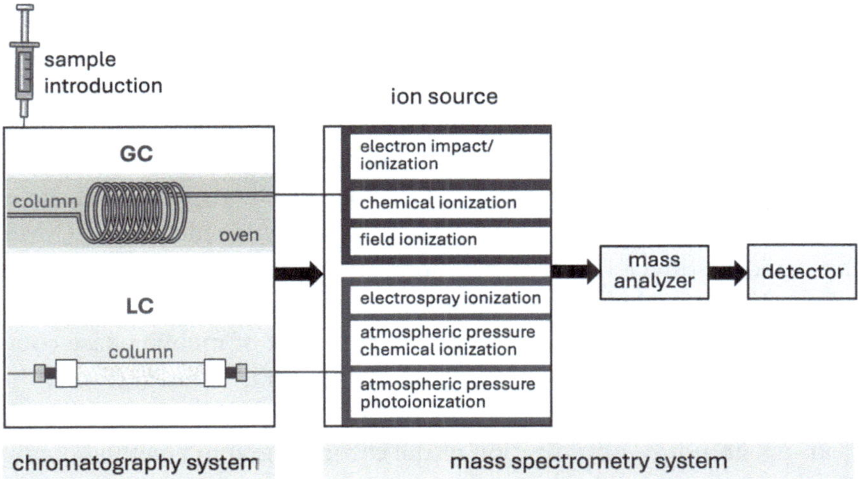

FIGURE 7.1 Schematic illustration of chromatography-MS systems.

ionization techniques, which can keep the molecular ions intact during the MS analysis. EI with lower energies can also be opted for to reduce the degree of fragmentation. In CI, ions are produced via proton addition $[M+H]^+$ or abstraction $[M-H]^-$ by reacting the gaseous molecules with excess reagent gas such as methane, ammonia, and isobutane [45]. FI produces molecular ions in the form of radical cations $[M^{+\bullet}]$, like those produced by EI, by applying a highly positive potential to an electrode with a sharp point, causing electron tunneling. FI-MS is often used to complement EI-MS to reliably assign the molecular mass of unknown compounds whose molecular ions are absent in EI spectra [46]. Mass analyzers such as quadrupole (most common), ion trap, TOF, and magnetic sector can be used in GC-MS [47].

LC separates components of a sample due to the differential affinity of the compounds toward the stationary phase, which can be solid particles or liquid coated onto solid supports uniformly packed in a column. Eluted components then enter the MS interface and are converted into ions via ESI, atmospheric pressure chemical ionization (APCI), or atmospheric pressure photo-ionization (APPI). LC-ESI-MS is the most common, in which a high voltage is applied to the liquid to form charged microdroplets. ESI produces multiply charged ions for large molecules such as proteins, dental resins, and other polymers, allowing analysis at lower *m*/*z* values [34, 48, 49]. ESI is mainly used for ionic species or compounds that are easily ionizable [50]. APCI is similar to CI, in which reagent gases are used to produce molecular ions. However, APCI is operated at atmospheric pressure, enabling a higher frequency of collisions between the solvated analytes and reagent molecules. APCI and APPI are more efficient in ionizing nonpolar compounds [50]. In APPI, analytes are ionized using a vacuum ultraviolet light source operating at atmospheric pressure. This ionization technique can produce higher signals (more sensitive) than APCI [51]. Although the matrix effect has been reduced in LC-MS, compared to direct injection MS, ionization suppression/enhancement may still occur due to co-elution with interfering compounds, co-precipitation, ionization competition, or influence of mobile phase components [52]. To address this problem, several strategies can be implemented, such as selective extraction of analyte(s) of interest, sample clean-up, and/or re-optimization of the chromatographic method/conditions to achieve complete separation of the analytes from other components in the sample matrix. Internal standards can also be incorporated for reliable quantification using LC-MS.

ESI, APCI, and APPI are all soft ionization techniques that can preserve the molecular integrity of the analytes. For reliable analyte identification, tandem MS or MS/MS is often needed to provide a combination of information on accurate masses of the ions and fragmentation patterns. LC-tandem MS has been used for both identification and quantification in dental materials/samples using several similar mass analyzers such as triple-quadrupole [41, 42, 53], or combinations of different mass analyzers such as quadrupole-linear ion trap [54], quadrupole-linear ion trap-Orbitrap [44], and quadrupole-TOF [55].

7.2.1 Application of Liquid Chromatography-Mass Spectrometry for the Analysis of Dental Materials and Related Compounds

Dental treatments for restorative dentistry increasingly use resin-based dental composites as dental restorative materials. These materials usually consist of a mixture of organic monomers as a resin matrix as the main component and filler composed of organic or inorganic materials [56]. Common materials for resin matrix are bisphenol A glycidyl methacrylate (BisGMA), bisphenol A dimethacrylate (Bis-DMA), urethane dimethacrylate (UDMA), and triethylene glycol dimethacrylate (TEGDMA). A dental composite usually contains several types of methacrylate monomers. In the presence of additive materials, including photo-initiator, co-initiator, and photostabilizer, the monomers are converted into polymers with a conversion of 43–70% [33]. Qualitative analysis of various ingredients usually used in dental materials has been successfully performed using liquid chromatography-quadrupole time-of-flight mass spectrometry (LC-QTOF-MS) to detect a total of 39 compounds [34]. A simpler instrumentation such as LC-MS is also suitable for qualitative analysis of the main monomers [32].

Most of the literature on the analysis of dental materials focuses on the quantification of monomers as the main component of dental materials. Unconverted monomers may leach and pose problems for dental health. Low conversion rate indicates inefficient reaction, resulting in lower quality of dental filling and monomer leaching, which can induce toxicity. Residual monomers were analyzed using liquid chromatography-tandem mass spectrometry (LC-MS/MS). External standard calibration was applied, but the multiple reaction monitoring (MRM) of the MS used Solvent Blue 35 (SB-35) as an internal standard. Ethanol/water (3:1) was used in the leaching study to simulate the environment in the oral cavity in a 28-day period. The leaching components from the dental materials depend on the manufacturer, but most materials release TEGDMA. Big monomers like BisGMA had lower concentration in the immersed solution [33].

Detection of leachable monomers in unstimulated saliva and artificial saliva has been done to simulate a more similar matrix to those in the patients. LC-MS/MS with negative and positive modes of ESI was used, applying MRM for the mass acquisition. The elution was performed in the C8 column using 10 mM ammonium acetate and methanol/water (95/5) containing 10 mM ammonium acetate as the mobile phase in gradient mode. The linearity achieved in the aforementioned matrix reached 0.999. The lowest and highest LOQs were shown by UDMA and HEMA, respectively, which were 1.5 ppb and 90 ppb, respectively. The accuracy was within the acceptable range of 70–130% in both unstimulated saliva (82–117%) and in artificial saliva (76–127%), except for BisEMA homologues (EO_2–4), which exceeded 130%. The precision measured at the LOQ level was <20% for all compounds [57].

Among different materials, BisGMA is currently the most popular material. Unfortunately, the degradation of BisGMA and other methacrylate monomers may produce bisphenol-A (BPA), known to have various toxicities, mainly to the reproductive organs. BPA may also present as a by-product during the synthesis of methacrylate monomers. Therefore, chemical analysis of dental composites using liquid chromatography often focuses on the quantification of BPA as a toxic material.

BPA is often found in the saliva due to the leaching of dental materials. An in vitro study proved that BPA increases estrogenic activity and triggers a wide range of metabolic disturbances [58]. Present in low concentration, an accurate and sensitive detection method is highly required for the quantification of BPA to assess the safety of materials over prolonged time. Early studies use high-performance liquid chromatography (HPLC) with UV/visible or fluorometer detector for the quantification of BPA in biological samples, such as saliva and blood. Reversed-phase column of C18 [59,60] or C8 [58] was used. Analysis of BPA using HPLC with a fluorescence detector, eluted with methanol/0.05 M phosphoric acid (55%/45% v/v), produces a detection limit of 100 ppb [59]. Lower detection limit of 5 ppb was achieved using HPLC with a fluorescence detector, eluted with acetonitrile/water (1:1). This analytical method achieved good selectivity, as dental material components of TEGDMA and Bis-DMA were not detected using a fluorescence detector but were detected using a photo-diode array (PDA) UV/visible detector with a higher detection limit of 200 ppb [60].

Liquid chromatography with an MS detector, such as LC-MS/MS, is now a popular instrumentation for BPA quantification. UV and fluorescence detectors often do not show enough selectivity to analyze BPA in the presence of Bis-GMA. LC-MS/MS with MRM for the MS/MS acquisition can achieve a detection limit of 0.06 ppb. However, this experiment was performed as an in vitro experiment, in which ethanol/water (3:1) was used to prepare standard solutions. External calibration was used for BPA quantification, achieving an r^2 of 0.9914 for the range of 0.2–6.4 ppb and an accuracy of 93–106% [53].

BPA elution in LC systems with MS detector is often problematic due to limited BPA ionization in an MS detector. To eliminate this problem, BPA and deuterated BPA-d_{16} as the internal standard were derivatized using pyridine-3-sulfonyl chloride to increase the ionization. The detection of BPA using ultraperformance liquid chromatography-tandem mass spectrometry (UPLC-MS/MS) provided good accuracy (96–101%) and precision (4–15% RSD) [61]. The quantification limit was about 206 ppb, which was much lower than similar method of the underivatized BPA, showing quantification limit of about 41 ppm [62].

7.2.2 Utilization of Gas Chromatography-Mass Spectrometry on Evaluation the Dental Materials and Related Substances

GC-MS has been utilized on both dental materials and chemical agents with applications in dentistry since the 1990s [9]. Since then, the utilization of GC-MS for dentistry research has been taking place prominently. As previously explained, in GC-MS, volatile samples are evaporated, carried by inert gas through the column, and eluted from the column in the order of increasing distribution ratio, which depends on the stationary phase, temperature, flow rate, amount of the sample, and the length of the column. For a conventional GC-MS, extraction of compounds from solid sample material is needed as liquid form of sample is required. In general, larger compounds with higher boiling points have longer retention. Typical columns used in GC are packed columns (spherical particle) and open tubular columns (mostly a coated inner wall of a tube/capillary). Capillary columns are less robust than packed columns, but working life can be extended by bonding the stationary phase onto the inner wall of the column. Till this day, capillary columns are most widely used. As for the MS detection, quadrupole MS with a mass-to-charge ratio (*m/z*) in the range of 25–600 *m/z* at an energy of 70 eV of electron ionization (EI) is commonly set for a typical GC-MS method. EI ionization mode used in the MS is a hard ionization technique that provides additional information on fragment ions of all the compounds detected to better assist high-confident identification.

One of the early utilizations of GC-MS was performed by Manabe, A., et al. (2000) in which they used GC-MS for the detection of trace amount of bisphenol A in dental materials due to its concern of xenoestrogenic properties [63]. Silylated extracts of dental materials were introduced to the GC system with apolar capillary column HP-1, using gradient temperature of 100–300 °C. Selected ion monitoring (SIM) MS was used to identify and quantify the compound in their study. As a result, BPA was detected in all unpolymerized dental materials tested. Using this method, BPA could successfully quantify in levels ranging from less than 1 µg/g to about 20 µg/g in commercial materials. This study also confirmed the release of BPA up to 91.4 ng BPA/g material after 24-h incubation in phosphate-buffered saline.

The GC-MS for quantification of organic eluates from polymerized resin-based dental restorative material had also been reported to evaluate its possible residual monomers, additives, and degradation products to quantify and separate the eluates [40]. CP-SIL 8 CB column was used with helium as the carrier gas at a constant flow rate of 1 ml/min. Splitless injection was applied at temperature of 250 °C. The oven temperature program started at 50 °C, increased at 50 °C/min to 120 °C (held for 5 min), then increased at 10 °C/min to 280 °C (held for 1 min). By using this system, they were able to quantify nine eluates: 2-hydroxyethyl methacrylate (HEMA), TEGDMA, hydroquinone monomethyl ether (MEHQ), camphorquinone (CQ), butylated hydroxytoluene (BHT), ethyl 4-(dimethylamino)benzoate (DMABEE), trimethylolpropane trimethacrylate (TMPTMA), oxybenzone (HMBP), and drometrizole (TIN P) leaching from specimens of four commonly used resin-based dental materials in ethanol and an aqueous solution. The level quantification of each substance was reached in micrograms mm$^{(-2)}$.

Methanolic leachates derived from polymerized materials were analyzed by means of GC-MS and were also published by Vouzara, T., et al. [64]. For quantitative purposes, they used caffeine at the *m/z* of 194, and then, each unknown peak was normalized to the caffeine peak. By this method, twenty different substances from the three resin-based materials were successfully revealed, with median value of the relative amounts for eluates measured in the range of 0.24–665.65%.

Despite the apparent benefit of conventional GC technique in dental materials analysis, the need for prior extraction as part of the sample preparation may lead to the loss of some amount of valuable compounds in the sample during the extraction process. This lengthy sample preparation requires more effort and time. Headspace gas chromatography (HS-GC) provides an alternative to GC by analyzing samples directly in the solid form with minimum sample preparation without extraction. In HS-GC-MS, the solid sample is directly put into an HS vial (typically 20 mL vial) and heated at a certain temperature. The forming gas accumulated in the vial is then injected onto the GC column. Alifui-Segbaya, F., et al. characterized dental methacrylates for chemical composition using HS-GC-MS [65]. In their study, photocured samples were frozen in liquid nitrogen, ground into powder, and directly put in the HS vial. As low as a 1 mL aliquot of headspace gas was injected onto the column with a total run time of only 15 minutes. They showed that, with post-illumination and the applied ethanol treatment, a substantial proportion of chemical compounds observed in the methacrylates decreased. However, other potentially toxic acrylic esters were observed in the absence of residual methacrylate. Another application was performed by Kostic M., et al., in 2021 where they utilized the combination of multiple headspace extraction analysis GC-MS and conventional GC to determine the content of residual monomer methyl methacrylate (MMA) in cold and hot polymerized poly(methyl methacrylate), with residual content of 15.75 mg_{MMA}/g_{PMMA} (in cold) and 10.96 mg_{MMA}/g_{PMMA} (in hot) [66].

Their method allows direct analysis of dental acrylic resin without prior sample preparation.

Another GC technique that has been utilized in dentistry research is inversed gas chromatography (IGC). It is a powerful technique widely used for measuring the physicochemical properties of solid surfaces, offering a simple, fast, and accurate technique. This method has been applied for material analysis in polymers, minerals, nanomaterials, surfactants, and pharmaceuticals. The main aspect of IGC that distinguishes it from conventional GC is the nature of mobile phase and stationary phase. In IGC, the sample of interest is placed into the column, being the stationary phase. Then, probe compound (appropriately selected volatile compounds with known physicochemical properties) is injected via an inert carrier gas through the stationary phase, and thus, it interacts with the sample under investigation. Then, adsorption and desorption equilibrium of the probe molecule on the stationary phase surface is established. The time of interaction of the probe molecule with the stationary phase and the shape of the obtained peak are used to characterize the sample material. One of the main parameters used to describe phenomena occurring on complex system is surface energy. IGC provides accurate way to assess changes in the nature of the surface that have undergone different modifications. More in-depth theory and instrumentation of IGC can be found in the review article by Mohammadi-Jam, S., and Waters, K.E., 2014 [67]. In dentistry research, IGC provides possibility to determine adhesion between materials, that is valuable in the examination of surface properties of dental composites. A study reported by Adamska, K. et al., showed that different values of the work of adhesion were observed according to the type of tooth tissue (dentin or enamel) by analyzing the surface properties of dental methacrylic resin with the addition of various zeolite fillers IGC [68]. They examined materials such as charisma, charisma classic, methacrylic resin without filler, and methacrylic composites containing filler. The materials were placed in glass, silanized columns; n-alkanes, heptane, octane, nonane, and decane were used as probe materials. To calculate the surface properties of the materials, physiochemical parameters of the probes were used. In this experiment, flame ionization detector (FID) was used instead of MS detector. Most of the applications of IGC for dental materials utilize FID; in fact, the utilization of MS detector for IGC has been very minimal. In 2017, Roman Galmadez, J., published a study of polymer-solvent diffusivity and solubility using IGC coupled to MS detector, but after that, to our knowledge, scientific reports on IGC-MS analysis have not been utilized more [69]. Despite the apparent benefit of MS detection to obtain molecular weight structural information through fragmentation, MS is not inheritably quantitative technique. Quantitative analysis with MS requires some consideration regarding the ionization and ion suppression probability of all compounds in the sample, which is most often not easy to estimate; thus, the addition of internal standard cannot be avoided. In this case, FID becomes a choice of more straightforward quantitative method as the area under the peak of a compound in the FID data is directly proportional to its quantity.

7.2.3 Advantages and Common Limitations of Chromatography-Mass Spectrometry Analysis for Dental Material

MS has emerged as an essential tool in evaluating contemporary dental materials, particularly those comprising intricate compounds like dental resins and their extracts. As previously mentioned, BPA becomes imperative to assess its presence in dental materials due to the ongoing safety concerns. In a study by Hope et al. (2016), BPA in dental materials was investigated using LC/UV and LC/MS/MS methods [70]. They observed that BPA concentrations measured with LC/UV were significantly higher compared to those obtained with LC/MS/MS, unfortunately attributing this disparity to the co-elution of an unidentified chemical, likely from a polymerization initiator during LC/UV analysis. Consequently, LC/MS/MS was deemed more accurate and selective for determining the actual BPA concentration in samples.

Research in dental materials has greatly benefited from a multi-method approach, especially since suitable reference materials are often unavailable [28]. By using MS method, it enables the evaluation of substances by matching their molecular weight and fragmentation patterns, allowing for the detection of trace levels [71]. Additionally, MS can also simultaneously analyze complex substances [34].

LC-MS and GC-MS are powerful analytical methods that revolutionize the identification and quantification of complex molecular compositions, enhance biomaterial assessment precision, and drive innovations in dental material design. Despite their versatility and sensitivity, these methods have limitations that affect their effectiveness in certain applications.

Matrix effects pose another challenge for MS analysis, particularly in complex dental materials containing both organic and inorganic components. Unknown substances in the sample can introduce unwanted noise, complicating the evaluation. Proper sample preparation is essential to mitigate these issues. This sample preparation will easily affect the MS sensitivity. A good and proper sample preparation sometimes can be time-consuming and error-prone. GC-MS analysis is limited to volatile and thermally stable samples. On the other hand, MS analysis can sometimes yield false-negative results due to incomplete ionization of compounds. Moreover, MS identification often relies on existing databases, and not all chromatographic peaks can be identified, highlighting the need for additional non-target screening workflows [34].

ToF Secondary Ion Mass Spectrometry (ToF-SIMS) is a qualitative technique for analyzing surface characteristics, providing information on molecular compounds and fragments of larger organic macromolecules from the sample's outermost surface [72]. ToF-SIMS combines high mass resolution and quantification capabilities but can analyze only a limited number of predefined ions simultaneously. Due to its high sensitivity, only high-purity solvents should be used [73]. This technique also has limited optical capabilities and can damage the sample surface during analysis, making it might be needed to complement X-ray photoelectron spectroscopy (XPS) [72].

7.3 CONCLUSION

MS, coupled with chromatography, has proven to be a highly effective method to be applied in several fields of study. It is considered a universal detector with various uses in many scientific areas, underscoring its importance in modern research and diagnostics. In dentistry, its role in the analysis and development of dental materials is indeed undoubtable. MS analysis has provided precise identification and qualification of dental materials, enabling the researchers to understand better the detailed composition. This information is essential to help further dental material development that provides safer, more effective, and longer-lasting.

Further advancements have made it the method of choice in developing new assays, with increased sensitivity, specificity, speed, and high-throughput chemical information above the existing traditional diagnostic techniques. Chromatography-MS has allowed the detection of the material at a trace level, which is helpful to prove the minor component that might have played an important role in performance effect, evaluate the safety of material use, and identify the potentially toxic substance. Its high-throughput capabilities allow to evaluation for several compositions in one running analysis. As the development of dental material will always continue, MS with chromatography is expected to remain a valuable tool in supporting this ongoing analysis. In the future, we are looking forward to seeing how chromatography-MS can play a bigger role in providing insight data to shape the next generation of dental material.

REFERENCES

[1] Paital B 2015 Mass spectrophotometry: An advanced technique in biomedical sciences *Adv Tech Biol Med* **4** 1–8.

[2] Thomson J J 1913 Bakerian Lecture: Rays of positive electricity *Proc R Soc Lond Ser A-Contain Pap Math Phys Character* **89** 1–20.

[3] Thomas J M 2006 J. J. Thomson: Winner of the Nobel Prize for Physics 1906 *Angew Chem Int Ed* **45** 6797–800.

[4] Aston F W 1922 *Isotopes* (London: E. Arnold & Company).

[5] McLafferty F W 2011 A century of progress in molecular mass spectrometry *Annu Rev Anal Chem (Palo Alto Calif)* **4** 1–22.

[6] Montgomery J, Muldner G, Cook G, Gledhill A, and Ellam R 2009 Isotope analysis of bone collagen and tooth enamel. In C. Lowe (Ed.), *'Clothing for the Soul Divine': Burials at the Tomb of St Ninian. Excavations at Whithorn Priory, 1957-67* (pp. 63–80) (Edinburg: Historic Scotland).

[7] Nixon G S 1969 Trace Element Content of the Hard Dental Tissues and Dental Plaque *Caries Res* **3** 60–74.

[8] Michelsen V B, Lygre H, Skålevik R, Tveit A B and Solheim E 2003 Identification of organic eluates from four polymer-based dental filling materials *Eur J Oral Sci* **111** 263–71.

[9] Pitt J J 2009 Principles and applications of liquid chromatography-mass spectrometry in clinical biochemistry *Clin Biochem Rev* **30** 19.

[10] Burmaster S, Smith R, Eick D, Kostoryz E L and Yourtee D 2002 In-vitro stability, metabolism, and transport of dental monomers made from Bisphenol A and Bisphenol F *Macromol. Biosci.* **2** 365–79.

[11] Berry T G, Summitt J B, Chung A K H and Osborne J W 1998 Amalgam at the new millennium *Journal of the American Dental Association* **129** 1547–56.

[12] Chin G, Chong J, Kluczewska A, Lau A, Gorjy S and Tennant M 2000 The environmental effects of dental amalgam *Aust Dent J* **45** 246–9.

[13] Mackey T K, Contreras J T and Liang B A 2014 The minamata convention on mercury: Attempting to address the global controversy of dental amalgam use and mercury waste disposal *Sci Total Environ* **472** 125–9.

[14] McLafferty F W, Todd P J, McGilvery D C and Baldwin M A 1980 High-Resolution Tandem Mass Spectrometer (MS/MS) of increased sensitivity and mass range *J Am Chem Soc* **102** 3360–3.

[15] Siqueira W L, Bakkal M, Xiao Y, Sutton J N and Mendes F M 2012 Quantitative proteomic analysis of the effect of fluoride on the acquired enamel pellicle *PLoS One* **7** e42204.

[16] Cabaña-Muñoz M E, Parmigiani-Izquierdo J M, Alonso F C and Merino J J 2019 Increased systemic malondialdehyde levels and decreased Mo/Co, Co/Fe2+ ratios in patients with long-term dental titanium implants and amalgams *Journal of Clinical Medicine* **8** 86.

[17] Salmon C R, Giorgetti A P O, Paes Leme A F, Domingues R R, Kolli T N, Foster B L and Nociti F H 2017 Microproteome of dentoalveolar tissues *Bone* **101** 219–29.
[18] Chakraborty P and Pradeep T 2019 The emerging interface of mass spectrometry with materials *NPG Asia Materials* **11** 1–22.
[19] Xian F, Hendrickson C L and Marshall A G 2012 High resolution mass spectrometry *Anal Chem* **84** 708–19.
[20] Fenn J B, Mann M, Meng C K, Wong S F and Whitehouse C M 1989 Electrospray ionization for mass spectrometry of large biomolecules *Science* **246** 64–71.
[21] Lapinska B, Rogowski J, Nowak J, Nissan J, Sokolowski J and Lukomska-Szymanska M 2019 Effect of surface cleaning regimen on glass ceramic bond strength *Molecules* **24** 389.
[22] Bayne S C, Ferracane J L, Marshall G W, Marshall S J and van Noort R 2019 The evolution of dental materials over the past century: Silver and gold to tooth color and beyond *J Dent Res* **98** 257–65.
[23] Gomez-Meda R, Esquivel J and Blatz M B 2021 The esthetic biological contour concept for implant restoration emergence profile design *J Esthet Restor Dent* **33** 173–84.
[24] Koutouzis T 2019 Implant-abutment connection as contributing factor to peri-implant diseases *Periodontol 2000* **81** 152–66.
[25] Shahi S, Özcan M, Maleki Dizaj S, Sharifi S, Al-Haj Husain N, Eftekhari A and Ahmadian E 2019 A review on potential toxicity of dental material and screening their biocompatibility *Toxicol Mech Methods* **29** 368–77.
[26] Meschi N, Patel B and Ruparel N B 2020 Material pulp cells and tissue interactions *J Endod* **46** S150–60.
[27] Zhang Y and Kelly J R 2017 Dental ceramics for restoration and metal veneering *Dent Clin North Am* **61** 797–819.
[28] Kaczmarek K, Leniart A, Lapinska B, Skrzypek S and Lukomska-Szymanska M 2021 Selected spectroscopic techniques for surface analysis of dental materials: A narrative review *Materials* **14** 2624.
[29] Tettamanti L, Cura F, Andrisani C, Bassi M A, Silvestre-Rangil J and Tagliabue A 2017 A new implant-abutment connection for bacterial microleakage prevention: an in vitro study *Oral Implantol (Rome)* **10** 172–80.
[30] Revilla-León M, Morillo J A, Att W and Özcan M 2021 Chemical composition, Knoop hardness, surface roughness, and adhesion aspects of additively manufactured dental interim materials *J Prosthodont* **30** 698–705.
[31] Revilla-León M, Al-Haj Husain N, Methani M M and Özcan M 2021 Chemical composition, surface roughness, and ceramic bond strength of additively manufactured cobalt-chromium dental alloys *J Prosthet Dent* **125** 825–31.
[32] Alshali R Z, Salim N A, Sung R, Satterthwaite J D and Silikas N 2015 Qualitative and quantitative characterization of monomers of uncured bulk-fill and conventional resin-composites using liquid chromatography/mass spectrometry *Dental Materials* **31** 711–20.
[33] Barišić M L, Sarajlija H, Klarić E, Knežević A, Sabol I and Pandurić V 2023 Detection of leachable components from conventional and dental bulk-fill

resin composites (high and low viscosity) Using Liquid Chromatography-Tandem Mass Spectrometry (LC-MS/MS) Method *Polymers (Basel)* **15** 627.

[34] Vervliet P, de Nys S, Boonen I, Duca R C, Elskens M, van Landuyt K L and Covaci A 2018 Qualitative analysis of dental material ingredients, composite resins and sealants using liquid chromatography coupled to quadrupole time of flight mass spectrometry *J Chromatogr A* **1576** 90–100.

[35] Volmer D and Jessome L L 2006 Ion suppression: a major concern in mass spectrometry *LCGC North America* **24** 498–510.

[36] Nanita S C and Kaldon L G 2016 Emerging flow injection mass spectrometry methods for high-throughput quantitative analysis *Anal Bioanal Chem* **408** 23–33.

[37] Michelsen V B, Moe G, Strøm M B, Jensen E and Lygre H 2008 Quantitative analysis of TEGDMA and HEMA eluted into saliva from two dental composites by use of GC/MS and tailor-made internal standards *Dental Materials* **24** 724–31.

[38] Durner J, Schrickel K, Watts D C and Ilie N 2015 Determination of homologous distributions of bisEMA dimethacrylates in bulk-fill resin-composites by GC–MS *Dental Materials* **31** 473–80.

[39] Deviot M, Lachaise I, Högg C, Durner J, Reichl F-X, Attal J-P and Dursun E 2018 Bisphenol A release from an orthodontic resin composite: A GC/MS and LC/MS study *Dental Materials* **34** 341–54.

[40] Michelsen V B, Moe G, Skålevik R, Jensen E and Lygre H 2007 Quantification of organic eluates from polymerized resin-based dental restorative materials by use of GC/MS *J Chromatogr B* **850** 83–91.

[41] Gonçalves A R, Do Nascimento H L, Duarte G H B, Simas R C, De Jesus Soares A, Eberlin M N and Marques L A 2016 Liquid chromatography-tandem mass spectrometry determination of p-chloroaniline in gel and aqueous chlorhexidine products used in dentistry *Chromatographia* **79** 841–9.

[42] Henkel K, Altenburger M J, Auwärter V and Neukamm M A 2018 Full validation of a method for the determination of drugs of abuse in non-mineralized dental biofilm using liquid chromatography-tandem mass spectrometry and application to postmortem samples *Talanta* **176** 360–6.

[43] Pujadas M, Pichini S, Civit E, Santamariña E, Perez K And De La Torre R 2007 A simple and reliable procedure for the determination of psychoactive drugs in oral fluid by gas chromatography–mass spectrometry *J Pharm Biomed Anal* **44** 594–601.

[44] Rexhaj F, Sabel N, Robertson A and Lundgren T 2023 Evaluation of method parameters for sound undecalcified dental enamel proteomics using liquid chromatography-mass spectrometry *Arch Oral Biol* **155** 105802.

[45] Sleeman R and Carter J F 2005 Mass spectrometry | Overview *in Encyclopedia of Analytical Science (Second edition)* ed P Worsfold, A Townshend and C Poole (Oxford: Elsevier) pp 337–44.

[46] Gross J H 2020 From the discovery of field ionization to field desorption and liquid injection field desorption/ionization-mass spectrometry—a journey from principles and applications to a glimpse into the future *Eur J Mass Spectrom* **26** 241–73.

[47] Maštovská K and Lehotay S J 2003 Practical approaches to fast gas chromatography–mass spectrometry *J Chromatogr A* **1000** 153–80.
[48] Koster S, Duursma M C, Boon J J and Heeren R M A 2000 Endgroup determination of synthetic polymers by electrospray ionization Fourier transform ion cyclotron resonance mass spectrometry *J Am Soc Mass Spectrom* **11** 536–43.
[49] Salmon C R, Tomazela D M, Ruiz K G S, Foster B L, Paes Leme A F, Sallum E A, Somerman M J and Nociti F H 2013 Proteomic analysis of human dental cementum and alveolar bone *J Proteomics* **91** 544–55.
[50] Ghislain T, Faure P and Michels R 2012 Detection and monitoring of PAH and Oxy-PAHs by high resolution mass spectrometry: Comparison of ESI, APCI and APPI source detection *J Am Soc Mass Spectrom* **23** 530–6.
[51] Robb D B, Covey T R and Bruins A P 2000 Atmospheric pressure photoionization: An ionization method for liquid chromatography–mass spectrometry *Anal Chem* **72** 3653–9.
[52] Furey A, Moriarty M, Bane V, Kinsella B and Lehane M 2013 Ion suppression; A critical review on causes, evaluation, prevention and applications *Talanta* **115** 104–22.
[53] Lopes-Rocha L, Gonçalves V M F, Cunha S C, Fernandes J O, Pinho T and Tiritan M E 2023 Evaluation of BPA and Bis-GMA release from recent dental composite materials by LC-MS/MS *Separations* **10** 455.
[54] Bathena S P, Huang J, Nunn M E, Miyamoto T, Parrish L C, Lang M S, Mc Vaney T P, Toews M L, Cerutis D R and Alnouti Y 2011 Quantitative determination of lysophosphatidic acids (LPAs) in human saliva and gingival crevicular fluid (GCF) by LC–MS/MS *J Pharm Biomed Anal* **56** 402–7.
[55] Chen C, Mondal K, Vervliet P, Covaci A, O'Brien E P, Rockne K J, Drummond J L and Hanley L 2023 Logistic regression analysis of LC-MS/MS data of monomers eluted from aged dental composites: A supervised machine-learning approach *Anal Chem* **95** 5205–13.
[56] Aminoroaya A, Neisiany R E, Khorasani S N, Panahi P, Das O, Madry H, Cucchiarini M and Ramakrishna S 2021 A review of dental composites: Challenges, chemistry aspects, filler influences, and future insights *Compos B Eng* **216** 108852.
[57] Grutle L A, Holm H V., Kopperud H B M and Uhlig S 2024 Validation of a human saliva model for the determination of leachable monomers and other chemicals from dental materials *J Chromatogr B* **1236** 124073.
[58] Olea N, Pulgar R, Pérez P, Olea-Serrano F, Rivas A, Novillo-Fertrell A, Pedraza V, Soto A M and Sonnenschein C 1996 Estrogenicity of resin-based composites and sealants used in dentistry. *Environ Health Perspect* **104** 298–305.
[59] Arenholt-Bindslev D, Breinholt V, Preiss A and Schmalz G 1999 Time-related bisphenol-A content and estrogenic activity in saliva samples collected in relation to placement of fissure sealants *Clin Oral Investig* **3** 120–5.
[60] Fung E Y K, Ewoldsen N O St. Germain H A, Marx D B, Miaw C-L, Siew C, Chou H-N, Gruninger S E and Meyer D M 2000 Pharmacokinetics of bisphenol a released from a dental sealant *J Am Dent Assoc* **131** 51–8.

[61] De Nys S, Putzeys E, Vervliet P, Covaci A, Boonen I, Elskens M, Vanoirbeek J, Godderis L, Van Meerbeek B, Van Landuyt K L and Duca R C 2018 A novel high sensitivity UPLC-MS/MS method for the evaluation of bisphenol A leaching from dental materials *Sci Rep* **8** 6981.
[62] Putzeys E, Cokic S M, Chong H, Smet M, Vanoirbeek J, Godderis L, Van Meerbeek B, Van Landuyt K L and Duca R C 2017 Simultaneous analysis of bisphenol A based compounds and other monomers leaching from resin-based dental materials by UHPLC-MS/MS *J Sep Sci* **40** 1063–75.
[63] Manabe A, Kaneko S, Numazawa S, Itoh K, Inoue M, Hisamitsu H, Sasa R and Yoshida T 2000 Detection of Bisphenol-A in dental materials by gas chromatography-mass spectrometry *Dent Mater J* **19** 75–86.
[64] Vouzara T, Roussou K, Nikolaidis A K, Tolidis K and Koulaouzidou E A 2020 Organic eluates derived from intermediate restorative dental materials *Molecules* **25** 1593.
[65] Alifui-Segbaya F, Bowman J, White A R, George R, Fidan I and Love R M 2020 Chemical characterization of additively manufactured methacrylates for dental devices *Addit Manuf* **31** 100944.
[66] Kostić M, Stanojević J, Tačić A, Gligorijević N, Nikolić L, Nikolić V, Igić M, Bradić Vasić M, Kosti M, Stanojevi J, Ta Ci A, Gligorijevi N, Sa Nikoli L, Nikoli V, Igi M and Bradi Vasi M 2020 Determination of residual monomer content in dental acrylic polymers and effect after tissues implantation *Biotechnol Biotechnol Equip* **34** 254–63.
[67] Mohammadi-Jam S and Waters K E 2014 Inverse gas chromatography applications: A review *Adv Colloid Interface Sci* **212** 21–44.
[68] Adamska K, Sandomierski M, Buchwald Z and Voelkel A 2020 Inverse gas chromatography in the examination of surface properties of experimental dental composites *Polym Test* **90** 106697.
[69] Román Galdámez J, Danner R P and Larry Duda J 2007 Application of mass spectrometer-inverse gas chromatography to study polymer-solvent diffusivity and solubility *J Chromatogr A* **1157** 399–407.
[70] Hope E, Reed D R and Moilanen L H 2016 Potential confounders of bisphenol-a analysis in dental materials *Dental Materials* **32** 961–7.
[71] Wang Q, Yu Q, Lin Q and Duan Y 2015 Emerging salivary biomarkers by mass spectrometry *Clin Chim Acta* **438** 214–21.
[72] Yahia L H and Mireles L K 2017 X-ray photoelectron spectroscopy (XPS) and time-of-flight secondary ion mass spectrometry (ToF SIMS) *in Characterization of Polymeric Biomaterials* ed Tanzi M C and Farè S (Oxford: Elsevier) pp 83–97.
[73] Al-Bataineh S A, Jasieniak M, Britcher L G and Griesser H J 2008 TOF-SIMS and principal component analysis characterization of the multilayer surface grafting of small molecules: Antibacterial furanones *Anal Chem* **80** 430–6.

CHAPTER 8

X-Ray Spectroscopy for Analysis of Dental Materials

Elif Sadik and Meryem Toraman

8.1 INTRODUCTION

Spectroscopy studies how atoms or molecules absorb, emit, or scatter electromagnetic radiation and is widely used to identify and analyze materials in various fields. Spectroscopic techniques can be classified such as nuclear, atomic, or molecular spectroscopy based on the type of electromagnetic radiation used in the analysis, such as x-ray, ultraviolet, or infrared, and the interaction between the matter and the energy, such as absorption, emission, or scattering/diffraction [1].

The electromagnetic spectrum is a range of frequencies of electromagnetic radiation, each corresponding to a specific wavelength and photon energy. X-rays within the electromagnetic spectrum between 10 and 0.01 nm have a variety of applications in diverse spectroscopic techniques [2]. Since its discovery by Wilhelm Conrad Röntgen in 1895, X-rays have had a wide range of applications, initially in medicine and subsequently in numerous scientific disciplines, including physics and chemistry. New applications for X-rays are continuously being researched and developed, expanding their utility. Material characterization using X-rays has undergone explosive growth since the early 20th century. Today, X-ray-based devices have become rapid analytical tools for characterizing materials across many scientific disciplines, including physics, chemistry, biology, archaeology, geology, astronomy, and material sciences [3].

DOI: 10.1201/9781003506805-8

X-ray spectrometers are instruments designed to examine X-ray beams to analyze incident X-rays' wavelengths, energies, and fluxes [4]. This method allows for the acquisition of detailed information about the composition and structure of materials. This technology emerged shortly after the discovery of X-rays and has made significant progress, especially with the groundbreaking work of Bragg and Bragg in 1913 [5], which provided the first detailed X-ray spectrum of crystal diffraction.

Various spectroscopic methods have been developed, and new devices have emerged that take advantage of the absorption, emission, or scattering interactions that occur when X-rays strike matter. The principles and applications of these methods exhibit considerable variability.

Dentists play a critical role in establishing and maintaining oral health. To achieve optimal oral health, it is essential to consider several factors, including the patient's needs and preferences, the expertise of the dentist, and the properties of the materials used. A comprehensive understanding of dental materials' physical, chemical, and biological properties is vital for maintaining oral and overall bodily health. The field of materials science is continuously advancing with developing technology. In recent decades, there have been rapid advances in the materials used for oral and dental health [6]. The dental materials market offers a wide range of products. With the advancement of technology, the number of available materials is increasing rapidly, and their properties are being improved daily.

Dental materials such as metals, ceramics, and polymers are applied directly to tissues and are closely related to health. Therefore, understanding dental materials' physical, chemical, and biological properties is crucial. X-ray spectrometric methods are frequently used for the analysis of dental materials. Commonly employed techniques include X-ray photoelectron spectroscopy (XPS), X-ray fluorescence spectroscopy (XRF), and X-ray diffraction (XRD) [7]. These methods provide valuable insight into the structure and composition of dental materials, helping to improve them and produce new materials.

8.1.1 X-Ray Photoelectron Spectroscopy

XPS, often called electron spectroscopy for chemical analysis (ESCA), is the most widely used non-destructive technique for surface characterization of materials. It is the most versatile due to its ability to be applied to diverse materials, providing invaluable quantitative and chemical state data from the surface of the material under study [8,9].

Surface analysis with XPS is performed by irradiating a sample with mono-energetic, soft X-rays and analyzing the energy of the electrons emitted from the sample [10]. The method is fundamentally established in the photoelectric effect elucidated by Einstein in 1905 [11]. The Einstein equation can describe the fundamental principles of this process:

$$E_B = h\nu - E_K$$

where E_B represents the binding energy, $h\nu$ is the energy of the incident photon, and E_K signifies the kinetic energy of the emitted electron, also referred to as a photoelectron, measured in the XPS spectrometer [9, 12]. In the 1950s, Kai Siegbahn irradiated atoms with photons and measured the energy of the emitted electrons using the photoelectric effect, developing methods for precisely measuring energy levels in atoms. In recognition of his invention, Siegbahn was awarded the Nobel Prize in Physics in 1981 [9,13].

The XPS device comprises four main components: the vacuum system, X-ray source, electron energy analyzer, and data system. XPS examinations are conducted in ultra-high vacuum conditions, typically ranging from 10^{-7} to 10^{-10} Pa (equivalent to 10^{-9} and 10^{-12} Torr). These conditions are essential to avoid energy loss due to collisions with gas molecules of the electron emitted from the sample, to ensure that the electrons reach the detector without scattering, and to maintain the sample surface uncontaminated by gas molecules [9,14]. The selection of anode material in XPS determines the energy of the produced X-ray transition; commonly used anode materials are aluminum (Al) and magnesium (Mg). These provide AlKα and MgKα photons with energies ($h\nu$) of 1486.6 eV and 1253.6 eV, respectively [15]. Soft X-rays, with an energy of approximately one keV, will penetrate 1000 nm or more into the analyzed material. In contrast, electrons of the same energy will only penetrate approximately 10 nm. For electrons to be detected, they must emerge from the material's surface. The free mean path inside the material limits the depth of analysis. In XPS, the electrons detected come from the material's top 5–10 nm, making XPS an inherently surface-sensitive analytical method [9,14]. A survey scan, also known as a wide-scan XPS, is performed with the binding energy ranging from 0 to 1200 eV in standard analysis. This method provides data on the elements contained in the sample. Narrow-scan XPS analysis can also focus on specific peaks to clarify the chemical state or environment [8].

XPS can detect and identify every element in the periodic table except hydrogen and helium [8, 9, 13, 14]. In analyzing a surface with XPS, the kinetic energy of the emitted photoelectrons is measured, and the spectrum is created. The resulting spectrum plots the number of electrons detected versus their kinetic energy. Because each element has a unique binding energy, it produces a unique spectrum with a characteristic set of peaks [16,17].

The number of electrons detected at each peak in a spectrum is directly proportional to the relative amount of the element within the XPS sample volume, allowing for quantitative analysis of the sample composition. The quantitative data is reported in atomic percent (%) [14]. XPS can detect elements at concentrations above 0.1% with high accuracy [9]. Furthermore, XPS can provide information on the chemical state, oxidation state, details of adjacent atoms, and electronic structure of elements, which is critical for understanding the surface chemistry of materials and studying processes such as adsorption, corrosion, and catalysis [14].

8.1.2 X-Ray Fluorescence Spectroscopy

X-ray fluorescence (XRF) is a non-destructive analytical technique based on X-ray emission spectroscopy. XRF analyzes the chemical composition of samples by detecting the characteristic secondary (fluorescent) X-rays emitted by samples after they have been irradiated with high-energy primary X-rays. This method is highly effective for rapidly analyzing numerous elements in powder, paste, solid, or liquid samples with remarkable sensitivity. Even when only minimal sample quantities are available, XRF can provide accurate and detailed elemental composition data [7,18].

An XRF device comprises three main components: the X-ray source, the detection system, and the data acquisition and processing system. The X-ray source produces primary X-rays that excite the atoms within the sample. When these primary X-rays irradiate the sample within the XRF device, the particles absorb the radiation, which can result in electrons being ejected from their inner shells. For the excited atoms to return to their ground state, the vacancy left in the inner shell is filled by an electron from an outer shell. During this process, characteristic X-rays are emitted. Each element, when excited, emits X-rays of characteristic wavelength (inversely related to the square of its atomic number) and intensity, which is roughly proportional to the amount of that element present. High atomic number elements emit short wavelength, high energy radiation, while low atomic number elements emit long wavelength, low energy radiation [19]. The characteristic X-rays emitted from the sample can be analyzed using their wavelengths or energies. Consequently, two distinct types of XRF exist: wavelength-dispersive spectroscopy (WDS) and energy-dispersive spectroscopy (EDS). XRF spectrometry was initially introduced as WDS in the early 1950s, with the EDS type developed years later. WDS has more complex instrumentation than EDS and offers higher resolution and a more comprehensive range of elemental analysis. Nevertheless, the EDS system offers advantages such as low cost and rapid analysis [20].

The XRF is employed for both qualitative and quantitative analysis [7,21]. The identification of elements is based on the wavelengths of the emitted X-rays (qualitative analysis). In contrast, the determination of the concentrations of these elements in the sample is based on the intensity of the X-rays (quantitative analysis) [18].

XRF spectrometers are available in various sizes, from handheld battery-powered field instruments to high-powered laboratory instruments. XRF spectrometers can function in air, helium, or vacuum atmospheres. The sensitivity of fluorescent X-ray detection is significantly influenced by the surrounding environmental conditions [22]. Vacuum conditions provide the best sensitivity, followed by helium, and air offers the lowest sensitivity. This difference is essential for the detection of light elements. For instance, an EDS-type spectrometer can detect elements down to carbon (atomic number 6) in a vacuum environment, sodium (atomic number 11) in a helium atmosphere, and aluminum (atomic number 13) in an air atmosphere [20].

8.1.3 X-Ray Diffraction

Solids are classified into two groups: amorphous and crystalline. In amorphous materials, such as glasses and many polymers, the atoms are arranged randomly, whereas, in crystalline materials, the atoms are arranged in a regular pattern and are repeated in three dimensions, with the smallest volume element defining the crystal. Approximately 95% of all solid materials are crystalline [23].

When a monochromatic X-ray beam strikes a sample, one of the phenomena is the production of X-rays scattered at the same wavelength as the incident beam. This type of scattering, also known as elastic or coherent scattering, leads to the phenomenon of diffraction. In 1912, Max von Laue reported that if X-rays had a wavelength similar to the spacing between atomic planes in a crystal, it was possible to diffract X-rays by a crystal and thereby obtain information about the arrangement of atoms in crystals. In 1913, Lawrence Bragg formulated Bragg's Law, which describes the diffraction of X-rays by atoms in crystals. Bragg's Law explains the diffraction of X-rays by a crystal lattice [24].

The incident X-rays strike the crystal planes with an incident angle θ (Theta) and diffract with an angle θ. Consequently, the angle between the incident and diffracted beams is always 2θ. This angle, known as the diffraction angle, is quantified in an experimental setting. The intensities and spatial distributions of the scattered X-rays form a specific diffraction pattern. Each crystalline solid has a unique XRD pattern as a unique fingerprint. Phases with the same chemical composition have different XRD patterns. A powerful non-destructive technique, XRD, analyzes crystal structure and size, identifies phases, and observes phase transformations [23].

8.2 X-RAY SPECTROSCOPY IN DENTISTRY

In recent years, there has been remarkable progress in dental materials, leading to a vast array of choices becoming available. However, this increase in material diversity also challenges dentists to select the most appropriate material. Since dental materials come into direct contact with tissues when applied to the oral cavity, it is essential to consider their physical, chemical, and biological properties.

One of the primary problems dental materials face is surface phenomena occurring at their interface with oral tissues and the environment. These phenomena are generally different from those observed in the bulk. They may result in the penetration of harmful substances into the bulk or leakage of substances from the bulk, which can alter the properties of the dental material and limit its lifetime [25].

Significant research efforts are devoted to creating or improving existing materials. For this purpose, appropriate methods are needed to investigate surface chemistry and physical phenomena in dental materials and adjacent dental structures. Surface phenomena occur in layers at various depths, ranging from a few nanometers (or less) to a few micrometers (or even more) on the sample's surface. Considering the importance of surface properties, XPS, the most valuable method for evaluating the surface layer, is frequently used to examine dental materials [25].

The XPS method has been employed to investigate a remarkable diversity of materials, including dental alloys, implants, protective coatings, ceramics, composites, glass ionomers, and polymers [25]. In a study conducted in 1976, the surface of tooth enamel treated with fluoride gel was analyzed using XPS, demonstrating that XPS could be used effectively to examine the surface of tooth enamel [26]. In the subsequent years, XPS was increasingly employed in investigations of dental materials and adjacent dental structures.

Dental implants' surface chemistry is a touchstone in achieving osseointegration and soft tissue compatibility. Therefore, various surface modification methods are applied to dental implants to improve their clinical performance [27]. XPS is a widely used method for surface analysis of dental implants with different structures [27–31]. Approximately half of the applications where dental materials are examined with XPS involve the study of dental implants [25].

Arys et al. [32] conducted an XPS analysis of titanium dental implants and surrounding biological tissues following osseointegration failure. The authors reported that the composition of the oxide layer on implants that failed osseointegration was significantly affected. The presence of zinc, sodium, copper, fluorine, sulfur, silicon, tin, and titanium was demonstrated in the surrounding tissue. Furthermore, XPS has been used to characterize implants with enhanced antibacterial activity by antibiotic decoration [33]. In biocompatible implants, XPS showed particular interest in evaluating the deposition and growth of hydroxyapatite on the titanium implant surface [34]. XPS has also provided evidence of calcium phosphate nucleation on titanium surfaces that may occur at the implant-bone interface [35].

Alloys are used in nearly every field of dentistry, with the market providing over 3000 unique alloys for various uses. These alloys are used in dental treatments, such as dental fillings, prostheses, implants, and orthodontic procedures. They are frequently in long-term direct or indirect contact with epithelium, connective tissue, or bone. Given the long-term close contact with vital tissue, it is critical to investigate and understand the biocompatibility of the alloy. Corrosion properties, which also affect other properties of dental alloys, such as aesthetics and strength, are of great importance in biocompatibility [36].

For years, most dental alloys were gold-based, meaning they contained gold as the most prevalent element. Nowadays, alloys can be based on silver, gold, nickel, cobalt, palladium, or titanium. Minor elements present in dental alloys can be even more diverse. More than 25 elements from the periodic table are used in today's dental alloys [36]. XPS is an optimal methodology for assessing the surface characteristics of dental alloys with a diverse array of elements.

Several scientific studies have employed the XPS method to examine dental amalgam's surface. In these studies, dental amalgam samples with different chemical compositions were stored in other environments, including air, water, and saline; samples were aged for various periods and subjected to different clinical finishing techniques (burnished and polished) [37–41]. Furthermore, the effect of titanium contact on the dental amalgam surface was also investigated [42].

A review of studies on spectroscopic methods used for surface characterization of dental materials between 2000 and 2020 revealed that X-ray spectroscopy methods were the most prevalent, and the number of studies conducted with these methods increased over the years [7].

8.3 CONCLUSION

X-ray spectroscopy, which provides essential information about materials' chemical and structural properties, is an invaluable tool in the study of dental materials. These techniques have the potential to revolutionize materials science. In the coming years, as has been the case so far, the contributions of X-ray spectroscopic methods will significantly benefit from developing new dental materials with advanced properties.

REFERENCES

[1] Pérez-Juste I and Nieto Faza O. 2015. Interaction of radiation with matter. *Structure Elucidation in Organic Chemistry: The Search for the Right Tools*, 1–26. Wiley.

[2] Wang B, Yang X, Chen S, Lu S, Zhao S, Qian Q, Cai W, Wang S and Zang Z. 2022. Flexible perovskite scintillators and detectors for X-ray detection. *Science*, 25(12):105593.

[3] Zimmermann P, Peredkov S, Abdala P M, DeBeer S, Tromp M, Müller C and van Bokhoven J A. 2020. Modern X-ray spectroscopy: XAS and XES in the laboratory. *Coordination Chemistry Reviews*, 423, 213466.

[4] Markert T H and Förster E. 2003. X-Ray spectroscopy. *Digital Encyclopedia of Applied Physics*: 89–113.

[5] Bragg W H and Bragg W L. 1913. The reflection of X-rays by crystals. Proceedings of the Royal Society of London. *Series A, Containing Papers of a Mathematical and Physical Character*, 88(605):428–438.

[6] Cui H, You Y, Cheng G W, Lan Z, Zou K L, Mai Q Y, Han Y H, Chen H, Zhao Y Y and Yu G T. 2023. Advanced materials and technologies for oral diseases. *Science and Technology of Advanced Materials*, 24(1):2156257.

[7] Kaczmarek K, Leniart A, Lapinska B, Skrzypek S and Lukomska-Szymanska M. 2021. Selected spectroscopic techniques for surface analysis of dental materials: A narrative review. *Materials*, 14(10):2624.

[8] Aziz M and Ismail A F. 2017. X-ray photoelectron spectroscopy (XPS). Membrane Characterization. Elsevier, 81–93.

[9] Ratner B D and Castner D G. 2009. Electron spectroscopy for chemical analysis. *Surface Analysis: The Principal Techniques*, 47–112. Wiley.

[10] Chastain J and King Jr R C. 1992. Handbook of X-ray photoelectron spectroscopy. *Perkin-Elmer Corporation*, 40:221.

[11] Einstein A. 1905. Über einen die Erzeugung und Verwandlung des Lichtes betreffenden heuristischen Gesichtspunkt. *Ann Phys* 17, 132–148.

[12] Casu M B and Chassé T. 2014. Photoelectron Spectroscopy Applications to Materials Science. *Handbook of Spectroscopy: Second, Enlarged Edition*, 1523–1556. Wiley.
[13] Hofmann S. 2012. *Auger-and X-ray photoelectron spectroscopy in materials science: a user-oriented guide* (Vol. 49). Springer Science and Business Media.
[14] Scopelliti M. 2021. X-Ray photoelectron spectroscopy. *Spectroscopy for Materials Characterization*, 351–382.
[15] Watts J F, Wolstenholme J. 2003. *An introduction to surface analysis by XPS and AES. Chapter 2*, 17–58. John Wiley & Sons.
[16] Bubert H, Rivière J C and Werner W S. 2011. X-Ray Photoelectron Spectroscopy (XPS). *Surface and Thin Film Analysis: A Compendium of Principles, Instrumentation, and Applications*, 7–41. Wiley.
[17] Shard A. G. 2020. X-ray photoelectron spectroscopy. In *Characterization of Nanoparticles*, Elsevier, 349–371.
[18] Verma H. R. 2007. X-ray fluorescence (XRF) and particle-induced X-ray emission (PIXE). *Atomic and Nuclear Analytical Methods: XRF, Mössbauer, XPS, NAA and B63Ion-Beam Spectroscopic Techniques*, 1–90.
[19] Karathanasis A D and Hajek B F 1996. Elemental analysis by X-ray fluorescence spectroscopy. Methods of Soil Analysis: Part 3 Chemical *Methods*, 5, 161–223.
[20] Leng Y. 2013. *Materials characterization: introduction to microscopic and spectroscopic methods*. John Wiley and Sons.
[21] Oyedotun T D T. 2018. X-ray fluorescence (XRF) in investigating the composition of earth materials: a review and an overview. *Geology, Ecology, and Landscapes*, 2(2), 148–154.
[22] Janssens, K. 2014. X-Ray Fluorescence Analysis. *Handbook of Spectroscopy: Second, Enlarged Edition*, 449–506.
[23] Sharma R, Bisen D P, Shukla U and Sharma B G. 2012. X-ray diffraction: a powerful method of characterizing nanomaterials. *Recent Research in Science and Technology*, 4(8), 77–79.
[24] He B B. 2018. *Two-dimensional X-ray Diffraction*. John Wiley and Sons.
[25] Preoteasa E A, Preoteasa E S, Suciu I and Bartok R N. 2018. Atomic and nuclear surface analysis methods for dental materials: A review. *AIMS Materials Science*, 5(4), 781–844.
[26] Hercules D M and Craig N L. 1976. Composition of fluoridated dental enamel studied by X-ray photoelectron spectroscopy (ESCA). *Journal of Dental Research*, 55(5), 829–835.
[27] Kang B S, Sul Y T, Oh S J, Lee H J and Albrektsson T. 2009. XPS, AES, and SEM analysis of recent dental implants. *Acta Biomaterialia*, 5(6), 2222–2229.
[28] Zinelis S, Thomas A, Syres K, Silikas N and Eliades G. 2010. Surface characterization of zirconia dental implants. *Dental Materials*, 26(4), 295–305.
[29] Korotin D M, Bartkowski S, Kurmaev E Z, Meumann M, Yakushina E B, Valiev R Z and Cholakh S O. 2011. Surface characterization of titanium implants treated in hydrofluoric acid. *Journal of Biomaterials and Nanobiotechnology*, 3, 87–91.

[30] Ameen A P, Short R D, Johns R and Schwach G. 1993. The surface analysis of implant materials. The surface composition of a titanium dental implant material. *Clinical Oral Implants Research*, 4(3), 144–150.
[31] Ausiello P, Bolli E, Kaciulis S, Gloria A, Lanzotti A, Martorelli M, Mezzi A, Montanari R, Richetta M and Varone A. 2023. Morphology and microchemistry study of three commercial dental implants. *Surface and Interface Analysis*, 55(6–7), 411–416.
[32] Arys A, Philippart C, Dourov N, He Y, Le Q T and Pireaux J J. 1998. Analysis of titanium dental implants after osseointegration failure: Combined histological, electron microscopy, and X-ray photoelectron spectroscopy approach. *Journal of Biomedical Materials Research*, 43(3), 300–312.
[33] He S, Zhou P, Wang L, Xiong X, Zhang Y, Deng Y and Wei S. 2014. Antibiotic decorated titanium with enhanced antibacterial activity through adhesive polydopamine for dental/bone implant. *Journal of The Royal Society Interface* 11: 20140169.
[34] Takadama H, Kim H M, Kokubo T and Nakamura T. 2001 XPS study of the apatite formation process on bioactive Ti-6Al-4V alloy in simulated body fluid. *Science and Technology of Advanced Materials* 2: 389–396.
[35] Combes C, Rey C and Frèche M. 1998. XPS and IR study of dicalcium phosphate dihydrate nucleation on titanium surfaces. *Colloids and Surfaces B* 11: 15–27.
[36] Wataha, J and Schmalz G. 2009. Dental Alloys. In: *Biocompatibility of Dental Materials*. Springer, Berlin, Heidelberg. 221–254.
[37] Hanawa T, Takahashi H, Ota M, Pinizzotto R F, Ferracane J L and Okabe T. 1987. Surface characterization of amalgams using X-ray photoelectron spectroscopy. *Journal of Dental Research*. 66(9): 1470–1478.
[38] Hanawa T, Gnade B E, Ferracane J L, Okabe T and Watari F. 1993. Compositions of surface layers formed on amalgams in air, water, and saline. *Dental Materials Journal*. 12(2), 118–126.
[39] McDermott A J, Kothari S, Short R D, Van Noort R and Alexander M R. 1998. Surface chemistry of a high-copper dental amalgam. *Journal of Dental Research*. 77(12), 1999–2004.
[40] Uo M, Berglund A, Cardenas J, Pohl L, Watari F, Bergman M and Sjöberg S. 2003. Surface analysis of dental amalgams by X-ray photoelectron spectroscopy and X-ray diffraction. *Dental Materials*, 19(7), 639–644.
[41] Talik E, Babiarz-Zdyb R and Dziedzic A. 2005. Chemical characterization of selected high copper dental amalgams using XPS and XRD techniques. *Journal of Alloys and Compounds*. 398(1–2), 276–282.
[42] Asami K, Takada Y and Okuno O. 2002. Adsorption of released ions from dental amalgams on titanium. *Surface and Interface Analysis*. 34(1), 144–147.

CHAPTER 9

Raman Spectrometry for Analysis of Dental Materials

Gülsün Akay

9.1 INTRODUCTION

Modern dentistry is focused more on the aesthetics of biomaterials due to the quick development of novel materials and their function in preserving and enhancing the health of the masticatory system [1]. Dental materials must include properties such as stability of mechanical properties, accuracy in imitating natural tooth structures, ease of processing during the application process, and biocompatibility [1–3]. Physical and chemical composition, surface structure, roughness, and reactivity are some key properties in evaluating a particular material for dental applications [2, 4, 5]. There are many different types of materials in dentistry, and dentists and dental technicians need to know the properties and limitations of these materials. The development of advanced spectroscopic and microscopic tools/techniques makes it possible to thoroughly examine the studied materials' chemical compositions and surface phenomena [1, 6].

The use of Raman spectroscopy for biomedical applications has recently increased due to the development of fiber optic probes. It advances in instrumentation, which has particularly increased research on structural chemistry and surface analysis of biomaterials [7].

This chapter aims to provide an overview of the examination of different dental materials' physiochemical and structural properties by imaging them with the Raman spectroscopy technique. This chapter will also discuss the application of the Raman spectroscopy technique, its basic

DOI: 10.1201/9781003506805-9

working principles, advantages and limitations, and methods of examining dental materials with Raman spectroscopy.

9.1.1 Principles of Technique

The technique was discovered by Sir Chandrasekhar Venkata Raman in 1928. The basis of the method is that when ordinary light (sunlight) passes through a pure medium (gas or water), a small amount of weak and scattered light is produced with a frequency different from the frequency of the incident beam [8]. When light hits a sample being examined, this phenomenon in which a small part of it scatters in several different directions and causes inelastic scattering is called Raman scattering. Raman spectroscopy is inelastic scattering spectra based on the Raman effect obtained using scattered light at a frequency different from the incident light. The phenomenon of elastic scattering of photons is called Rayleigh scattering [9].

Like infrared spectroscopy, Raman spectroscopy is a technique used to examine vibrational-rotational spectra [1]. Raman analyses typically use lasers with photon energies ranging from ultraviolet to near-infrared absorption.

The Raman spectrum can be obtained using two primary methods: Fourier transform (FT) and dispersion. The two methods differ in the wavelength of the incident radiation and the transform method [1]. Bands in Raman spectra are specific features of different molecular bonds of the sample under study and appear as a unique spectral "fingerprint" of each substance [10]. Each material specimen, compound, or chemical element examined provides a unique spectrum resulting from the excitation of vibrational modes of one or several different molecules [9]. Four main components make up the Raman spectrometer [1] (Figure 9.1)

- Laser source that can emit light, usually continuous wave (CW). The laser source can be easily focused on a small sample area. Lasers have different wavelengths, allowing examination at different penetration depths.
- Sample illumination and collection system.
- Wavelength selector: Since the intensity of Raman scattering is very weak, it is used between the sample and the detector to select only the relevant wavelength. The light passing through the wavelength selector is sent to a detector suitable for the type of laser used and

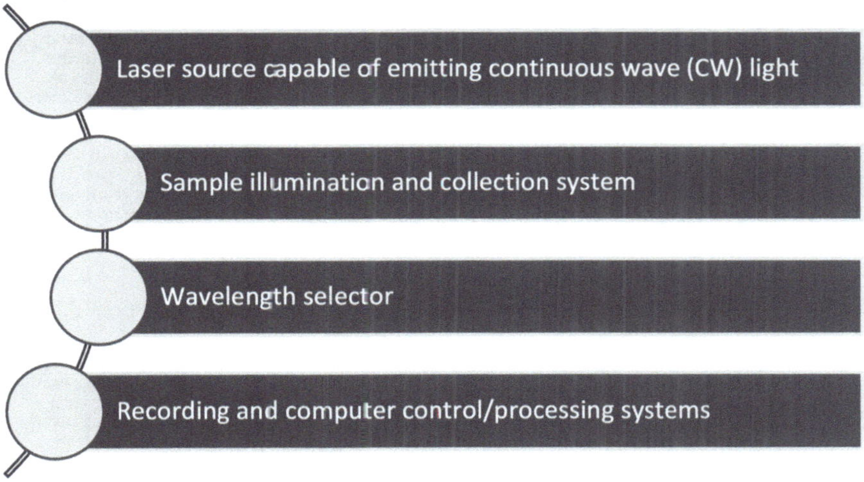

FIGURE 9.1 Fundamental components of Raman spectroscopy.

sensitive at that wavelength, where it is converted into an electrical signal and recorded.

- Recording and computer control/processing systems [1].

It can accurately and sensitively reflect changes in material structure and composition. Raman spectroscopy can be used to characterize materials, extracting properties such as irregularity, number of monolayers, interlayer breathing, thermal conductivity, strain and phonon modes, and in-plane anisotropy [1, 11, 12].

Raman spectroscopy can provide rich chemical information about the material under study, referred to as a molecular fingerprint. There are different approaches to analyzing the information content of the Raman spectrum. Some of those are as follows [13] (Figure 9.2).

9.1.2 Fields of Application

Raman spectroscopy is used to analyze solid, liquid, and gaseous samples, which can be organic, inorganic, or biological. Analysis of many water-soluble substances can be performed [12]. The Raman spectrum provides valuable information about the structure and chemical composition of the test substance [1]. The condensation process can study small amounts, and Raman spectroscopy now plays a significant role in studying biological systems and nanostructured materials under *in vivo* settings [14].

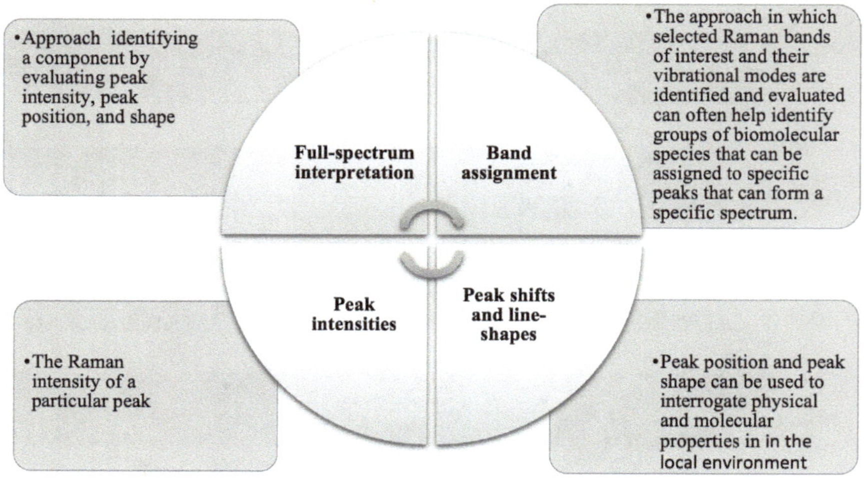

FIGURE 9.2 Methods for examining the information contained in a Raman spectrum.

Raman spectroscopy has become one of the most powerful tools in medical research due to advances and developments in technology, especially the development of new laser systems, CCD detectors, confocal Raman microscopy, and computer software [14, 15].

The Raman technique is suitable for analyzing biological samples because it is not affected by aqueous environments and is a scattering technique. It can simultaneously analyze carbohydrates, lipids, proteins, fatty acids, amino acids, nucleic acids (DNA analysis), and polysaccharides in a single spectrum with minimal sample preparation [15].

Raman spectroscopy is widely used in studying proteins along with other vibrational spectroscopy techniques, such as Fourier transform infrared spectroscopy (FTIR) [7]. Raman spectroscopy is a very suitable method for analyzing inorganic superficial texture (carbonate and phosphate) and observing the characteristic energies of the vibrational modes of the minerals present [9].

Vibrational spectroscopic techniques, such as Raman spectroscopy, are used to analyze biological tissues [7]. Raman imaging techniques can be used to characterize bone tissue, determine its general composition, and determine the spatial distribution of minerals in osteochondral tissue [13].

Various disease biomarkers can be identified in Raman spectra that can be used to diagnose or monitor the progression of certain disorder conditions [16]. Biochemical changes of biomarkers in malignant tumors on a cellular basis are detected in the vibration spectrum, and a comparison is made with the basic spectrum in normal tissue. Raman spectroscopy has proven specific and sensitive in the biomedical field [17].

With 96% inorganic substance, tooth enamel is the most mineralized structure in the body. Calcium phosphate ($Ca_{10}(PO_4)_6(OH)_2$) and hexagonal hydroxyapatite make up the inorganic material's chemical composition. The mineral phase was also described using Raman spectroscopy [7].

Recent studies have obtained promising results with Raman spectroscopy in the diagnosis of early dental caries and helped characterize hydroxyapatite crystals [17–19]. In Raman spectroscopy studies, the hydroxyapatite structure in enamel or dentin was characterized [7]. Raman spectroscopy has also been shown to be used in applications such as detecting early caries and caries-related bacteria, determining defective dentin margins, investigating the remineralization effect of drugs and the effects of radiation, and evaluating new bonding systems [20]. Studies have observed that the anisotropy of Raman spectra decreases due to demineralization during the caries process [21].

In a study using micro-Raman spectroscopy to examine caries-affected dentin, single-point detection was used to distinguish between healthy and caries-affected dentin. The study showed that carious dentin exhibited significantly smaller phosphate and carbonate/collagen ratios [22]. Raman spectroscopy has also been used in studies to evaluate the similarity between natural and regenerated dentin based on chemical composition and to examine dentin regeneration [23].

9.1.3 Advantages/Limitations

In Raman spectroscopy, spectral analysis is performed in reflection mode, and tissues can be investigated in their natural state without any or minimal sample preparation. This technique has advantages such as short measurement time, simple sample preparation, simple band and spectral analysis, high sensitivity to minor structural changes, and un-damaged sample analysis [24]. Additionally, the technique is non-invasive, and the results obtained can be analyzed easily and quickly. It does not show any sensitivity depending on the condition of the material/sample examined but is very sensitive to morphological, chemical, and structural changes. It also provides the opportunity to perform in vivo analysis [9].

When Raman spectroscopy began to be used in the biomedical field, it had some disadvantages, such as the lack of sensitive instruments and fluorescence from biological tissues. However, with the advancement of this technology, spectral analysis is performed in reflection mode, offering the opportunity to examine tissues in their natural state with little or no sample preparation [7, 17]. Some of its limitations include the high cost of fiber optic-based technology and benchtop equipment, the need for frequently sophisticated multivariate data analysis algorithms, low scattering efficiency that lengthens measurement times, shallow population depth of the light beam, and the fact that it cannot be used with pure metals and alloys. It may cause irreversible severe damage in vivo due to the heating of the sample (it may denature and decompose depending on the power used and acquisition time) [9]. Its shortcomings include low signal-to-noise ratio, weak signal, and self-fluorescence solid background [20].

9.2 ANALYSES OF DENTAL MATERIALS

With a wide variety of dental materials (ceramic, metallic, synthetic, or composite) available in dental practice, new materials, research techniques, and technological processes are constantly developed. However, factors such as physical, chemical, and biological properties and biocompatibility of the dental material in contact with the oral cavity play an important role. Accuracy in imitating natural teeth and maintaining the stability of their aesthetic and mechanical structure are other features that must be met [1]. To be successfully applied in the clinic, medical materials must be non-toxic, biocompatible, and possess a specific set of properties, depending on these materials' structural and molecular properties [13].

The advanced development of microscopic and spectroscopic techniques allows detailed examination of the studied materials' surface properties and chemical structure [6]. Vibration spectroscopic techniques provide in-depth information about components' molecular structure, dynamics, and interactions [13]. Raman spectroscopy offers both chemical and physical information about the specimen examined, such as the number of monolayers, in-plane anisotropy, and thermal conductivity [16].

It is a valuable tool for assessing the bio-absorption and biocompatibility of biomedical materials, quality control, and comparison [13]. Raman spectroscopy is the preferred method for precisely identifying the functional group of chemical compounds, especially organic compounds, and their interactions with human dentin and enamel [7].

The chemical process that causes the materials to change from a fluid to a solid state is one of the key components of the qualitative and quantitative study of dental materials, which evaluates the materials' condition both during and after hardening. Chemical reactions that occur with bond formation and breaking in materials are analyzed with spectroscopic techniques [11, 25]. Raman spectral analysis can also be used to assess the material's composition, stress/strain state, and symmetry or orientation of the inorganic filler particles [7]. Additionally, the change in polarization of the molecules within the material is assessed using Raman spectroscopy [11].

Today's most often utilized dental materials include mineral bone graft materials, ceramics, acrylic resins (polymethyl methacrylate), composite resins, metals (cobalt, nickel, and titanium), and implant materials. Raman spectroscopy is used in the literature to determine the polymerization rate and degree of transformation of various materials such as bioglass/composite/ceramic, to identify and test the presence of multiple materials such as ceramics and implant materials, to analyze and evaluate the phase composition of compounds, tooth tissue/material interfaces, as well as sample preparation, used to describe the effect of light-curing regimes and storage conditions [26–32].

9.2.1 Dental Implant Materials

Dental implants are a method used to restore chewing, phonation, and aesthetic capacity. The success of dental implants depends on the close contact between the implant and the bone. During the healing process after implant surgery, the immature bone is replaced by a mature type of bone containing inorganic components, most notably calcium hydroxyapatite (CHA [$Ca_{10}(PO_4)_6(OH)_2$] [33].

Raman spectroscopy is preferred in research for the analysis of TiO_2 phases on the root surfaces of dental implants. It was applied to assess the TiO_2 phase and the composition present on the implant surface [34]. Unlike rutile or anatase-rich TiO_2 surfaces, hydroxyapatite does not develop easily on amorphous TiO_2. Consequently, TiO_2 content affects an implant's chemical stability and longevity [34].

Lupi et al. evaluated the presence and transformation of TiO_2 on the implant surface at different production stages through micro-Raman spectroscopy. The results of their study were that as the complexity of the surface treatment increased, it caused a proportional increase in the thickness of amorphous TiO_2 oxide. Still, no crystalline TiO_2 formation was observed [35].

Titanium and its alloys (Ti-6Al-4V) have been used as dental implant materials due to their natural bone integration ability and high mechanical properties. In a study, graphene nanocoating covering Biomedical Grade Ti-6Al-4V alloys was characterized by Raman spectroscopy [36]. The results of their study confirmed high structural stability and resistance to mechanical stress and chemical degradation, retaining >99% coverage after corrosion [36].

Studies also evaluated bone healing around dental implants using Raman spectroscopy. In their study investigating the bone healing level around dental implants (with or without low-level laser therapy), Lopes et al. observed that there was a significant difference in calcium hydroxyapatite (CHA; ~960 cm^{-1}) between both groups. They concluded that bone healing processes can be safely evaluated with Raman spectroscopy [33].

9.2.2 Bioceramic Materials

Bioceramic materials constitute a significant part of inorganic and nonmetallic structures. Bioceramics have properties such as bioactivity, biocompatibility, durability, osteointegration, and tissue regeneration. They are also materials used in dentistry applications such as implantology, restorative, and endodontic treatments [37].

Some bioceramics commonly used in dentistry are hydroxyapatite, tricalcium phosphate, and bioactive glasses [37]. Hydroxyapatite is one of the bioceramic materials, and its chemical similarity to bone, biocompatibility, and being a bioactive and stable material are some of its features. To promote the attachment of implants to bone tissue, hydroxyapatite is typically employed as an implant coating [34].

In a study conducted with Fourier-transform Raman microscopy to evaluate the osseointegration of hydroxyapatite-coated implants, Raman spectra were used for the quantitative determination of inorganic and protein components. The study results showed that the Raman microscope can quantitatively analyze the composition of inorganic and organic components at the implant/bone interface [27].

9.2.3 Ceramics

Dental ceramics are preferred materials for various aesthetic restorations due to their properties, such as high compressive strength and wear resistance, biocompatibility, chemical stability, thermal expansion coefficient like tooth structure, and aesthetic properties [38]. The material's properties depend on the filler particles' type, size, distribution, and concentration. The most common filler ingredients are quartz, silica, and glasses such as boro-silicate, alumino-silicate, and barium oxide [11].

Dental ceramics can be divided into four main classes. These are silica (SiO_2 based), alumina (Al_2O_3 based), zirconium (ZrO_2 based) and calcium phosphate ceramics. It can form chemical bonds with living tissue by showing bioactive properties in some ceramics (bioglass) [39].

The content of dental porcelain is 3–5% kaolin (clay), 12–22% quartz (silica, sand), and 75–85% feldspar (mixture of potassium alumina silicate and sodium alumina silicate) [40]. Raman microspectroscopy investigated structural changes in silicate and aluminosilicate glasses. In a study in which the chemical transformations of porcelain kept in a simulated oral environment were analyzed by Raman spectroscopy, it was reported that the porcelain structure was significantly depolymerized by acid, causing the formation of ordered boron-containing species near the porcelain surface [41].

de Carvalho et al. evaluated the chemical composition of six different ceramic systems using Raman spectroscopy. The study results show that each ceramic system has different characteristic vibration modes, AL (Vitreous ceramic infiltrated by alumina: SiO_2 (0.04%); Al_2O_3 [99.8%]) and ALYZ (Alumina infiltrated vitreous ceramic and zirconia stabilizing particles: Al_2O_3 (64.8%); ZrO_2 [35%]) samples, it was observed that the appearance of vibration bands specific to ceramic materials was prevented by the fluorescence emission resulting from high Al_2O_3 concentration [38].

9.2.4 Zirconia Material

With its mechanical properties, biocompatibility, and aesthetic aspects, zirconia has become frequently used in crown-bridge prostheses. This ceramic has a phase transformation, a process responsible for the material's fracture toughness and high mechanical resistance [42]. Zirconia is a polymorphic material that exists in three different phases: monoclinic (m), tetragonal (t), and cubic (c). The mechanical characteristics, transformation process, and kinetics of zirconium were examined in the conducted research. Hydrothermal stress has been documented to cause material aging. Microcracking arises because of the increased volume and concentration of stress during the phase shift of the tetragonal structure to the monoclinic grain. Due to the small spatial resolution achieved by Raman spectroscopy (in microprobe and confocal configuration), it has become an important tool in studying the local analysis of the T-m phase transformation. Different equations have been used in the literature to determine monoclinic and tetragonal phase change and the composition of mixtures [43].

Band relative intensity, position, or width from Raman spectra can be used to carry out quantitative analyses. Raman sampling depth in zirconia ceramics (ZrO_2) can range from 5 μm to more than 50 μm depending on wavelength, objective lens, and laser intensity, among other variables [44].

Wulfman et al. published a study on measuring the transformed volume of zirconium material after in vitro aging since tetragonal and monoclinic zirconia phases have characteristics and different Raman spectra. The study results showed that Raman spectroscopy allows the examination of the transformation mechanism and diffusion into a large part of the material [42].

9.2.5 Resin-Based Composites

Restoration of dental caries demands specialized materials that are easy to model, naturally colored, light-cured, and possess clinical qualities like no leakage, less shrinkage, and hardness like hydroxyapatite. Resin-based composites are currently used as dental restorative materials due to their aesthetic and mechanical properties [26]. However, one of its most important limitations is polymerization shrinkage. Due to shrinkage, marginal gaps occur between the enamel and the composite, which causes marginal leakage and postoperative sensitivity [45]. The structural properties of composites were strengthened with the development of methacrylate-based monomers (Bis-GMA: bisphenol-A diglycidyl dimethacrylate), inorganic filler contents, silane coupling agents, and photo-initiator systems. Some initiators are added to the mixture to initiate the polymerization process of the light-cured material (especially due to reaction kinetics). The most used photo-initiator system consists of camphorquinone (CQ) and tertiary amine, and under visible light irradiation at a wavelength of approximately 450 nm, the C=C bonds of the resin monomers are converted into C—C bonds, forming high-strength, cross-linked polymers [26].

Raman scattering spectrum maps can be used to examine the distribution of composite components, mainly inorganic fillers, in the organic matrix and to determine the change in the degree of transformation on the examined surface [39]. Resin-based composite is characterized by the degree of transformation, which depends on the curing time and affects the quality of the filler [39].

Gatin et al. used Raman spectroscopy to calculate the degree of conversion (DC) (the number of π carbon bonds converted to σ ones) of nano- and micro-photopolymerizable dental composites. The study results showed a DC of 59% for sample 1 (Bis-GMA and TEGDMA resins, microcomposite) and 93% for sample 2 (TEGDMA, BIS-EMA resin, nanocomposite) after 45 seconds of curing [45].

Gomes de Araújo-Neto et al. investigated the mechanical and physical properties of seven different composites in their study published in 2021. Their study evaluated the degree of transformation and maximum rate of polymerization of composites using micro-Raman spectroscopy. Their results showed that the conversion degrees of all tested composites were between 78.5% and 92.8% [26].

A study comparing the photopolymerization process of three distinct resin-based composite materials, measurements of microhardness and density, the DC of monomer, and the material's stability using Raman spectroscopy found that the three composites' physical properties differed significantly. Furthermore, increasing the monomer concentration by 41% in a sample resulted in a decrease in the degree of curing from 74.8% (at a 5-second curing time) to 72% (at a 15-second curing time) and increased to 80.4% (at a 20-second curing time). It was determined that it increased [46].

9.2.6 Fiber-Reinforced Composites

Fiber-reinforced composites, which have good biomechanical properties, the advantage of minimally invasive restorations, and the ability to bond to teeth, have recently begun to attract attention in dental applications [11]. It is used in fixed prosthetic teeth, endodontic posts, splinting of mobile teeth, and composite veneers [47].

Fiber-reinforced composites are generally bis-GMA-PMMA (dimethacrylate-monomethacrylate) polymer matrices containing 1 wt% camphorquinone and DMAEMA (N,N-dimethyl aminoethyl methacrylate) as photo-initiator. Its main advantages are high tensile strength, chemical resistance, high sensitivity to abrasion, and low costs [11, 48].

Biocompatibility issues related to its chemical composition and stability may arise when fiber-reinforced composites are in contact with living tissues. In the studies of Bonon et al., the physicochemical properties and degradation profiles of three different fiber-reinforced composites (epoxide-based) materials were evaluated by various methods, including characterization, Brunauer–Emmett–Teller (BET) analysis, FTIR, Raman, X-ray fluorescence spectroscopy, and degradation experiments. The study's findings demonstrated that the laser beam's dispersion/scattering by the glass fiber caused a significant rise in baseline between 1000 and

2500 cm^{-1}, and that a distinct band indicative of C–C stretching was clearly seen at 1600 cm^{-1} [49].

9.2.7 Dentine Bonding Agents/Adhesives

Dental adhesive systems, which establish an effective bond with dental structures, are widely used in clinical applications. Effective bonding depends on the adhesive's chemical composition, the material's clinically convenient use, and knowledge of the morphological changes in dental tissue caused by different bonding procedures [11]. Dental adhesives can be classified into two subgroups, "self-etch adhesives" and "etch-and-rinse adhesives", according to how they interact with tooth structures in different clinical applications. Etch-and-rinse systems first pre-treat dental hard tissues with phosphoric acid and then rinse. Self-etch adhesives consist of an acidic monomer that simultaneously erodes and primes the tissue [50].

Raman spectroscopy can detect changes in the chemical compounds of the dentin-bonding adhesive complex. Santini et al. examined the degree of demineralization of dentin, adhesive penetration, and DC of self-etch adhesives (total etch, one-step self-etch, two-step self-etch) along the dentin-adhesive interface with micro-Raman spectroscopy. The study results reported statistically significant differences in dentin demineralization depth and adhesive penetration among all adhesives. Additionally, a significant number of unreacted monomers were detected in all systems, especially in the etching and rinsing systems [51].

In their study, Navarra et al. examined the DC of silorane-based adhesive using micro-Raman spectroscopy. When they compared silorane-based adhesives with methacrylate-based adhesives, they observed that silorane-based adhesives reached very high conversion degree values [52].

Miyazaki et al. examined the chemical content/component of the resin-impregnated dentin layer with micro-Raman spectroscopy in their study, and the findings of the study determined that the dentin-resin layer was not homogeneous and that the adhesive resin had poor fullness with the one-bottle adhesive system [53].

9.2.8 Glass Ionomer Cement

Glass-poly(alkenoate) dental cement, known as glass ionomer cement (GICs), is set by an acid-base reaction between an acid-degradable glass and a polymeric acid (polyacrylic acid [PAA]). Modified glass ionomer cements (RMGICs) exhibit activity due to free radical polymerization following photocatalyzation of a hydrophilic, methacrylate-based monomer.

It is essentially a cement with surface qualities that cosmetically resemble natural teeth, is highly biocompatible, is insoluble in water, and binds to the tooth with strong forces and minimal shrinkage. The degree of GIC hardening is influenced by sample solubility, cement characteristics (such as pressure and bending), and surface hardness [54].

Raman spectroscopy reflects the weak scattering effects of water and glass (with Si-O bonds). Acid-base and polymerization reactions in resin-modified GIC can be more easily measured by the Raman spectrum than FTIR [11].

In the investigation of Young et al., the Raman spectra of various combinations of components of GIC were compared with those of the reactants and tartaric acids and salts of polyacrylic. As a result of the study, changes in Raman spectra were determined during the reaction between very fast-hardening, acid-degradable fluoro-alumino silicate glass, polyacrylic acid, tartaric acid, and water at different acid and water ratios [54].

9.2.9 Fissure Sealant

Tooth decay is one of the most common and preventable conditions in children, and in clinical practices, fissure sealants are placed in the pits and fissures of the teeth. Thus, it connects to the tooth micromechanically and reduces caries risk [11]. Resin-based fissure sealants contain resin monomers like resin-based composites, and their use is widespread. The composition of the material, reaction kinetics, and monomer-to-polymer conversion are important for optimum mechanical properties and resistance to resin-based biodegradation [51].

In their investigation, Santini et al. assessed the potential effects of two distinct polymerization lights (a light-emitting diode, or LED, and a quartz-tungsten halogen light, or QTH LCU) on the conversion degrees of three distinct opaque and transparent fissure sealants, as determined by micro-Raman spectroscopy. Demonstrated that, for all three materials, LED generated much greater conversion degree values than QTH and that surface conversion degree values were significantly higher than subsurface values [51].

9.2.10 Zinc Phosphate Cement

With its good mechanical properties and minimal water solubility, zinc phosphate cement has been used as a temporary filling material and a lining material under restorations such as amalgam fillings. It is obtained by mixing the structure consisting of powder (zinc oxide – 90% and other metallic oxides – 10%) and liquid (50–60% phosphoric acid solution and buffers such as $AlPO_4$ and $Zn_3(PO_4)_2$ up to 10%) [11].

Zinc phosphate cement is fixed to the restorations completely mechanically. Its compressive strength is generally good, and its tensile strength is low. It undergoes dissolution/chemical changes in the oral environment [55]. Raman spectroscopies are a sensitive technique that can provide information about the chemical composition of zinc phosphate cement. In the examination of zinc phosphate minerals by Raman spectroscopy, minerals can be characterized by patterns of hydroxyl stretching vibrations in the spectrum [56]. By placing zinc phosphate cement into simulated body fluid and artificial saliva, its phase composition was examined in a study by Raman spectroscopy. While no change was observed in the peaks in the spectrum of the sample incubated in simulated body fluid, it was observed that the spectrum of the sample incubated in artificial saliva included the intense band at 1000 cm^{-1}, which can be assigned to the stretching vibrations of the $CaHPO_4 \cdot 2H_2O$ and $Zn_3 (PO_4)_2.4H_2O$'daki PO_4^{3-} group [57].

9.2.11 Zinc Oxide Eugenol Cement

Zinc oxide (ZnO) eugenol is used as a temporary filling material, base and lining, and luting cement due to its antibacterial and sedative properties and good insulation. Typically, 80–90% of its substance is made up of ZnO powder, with eugenol-binding resin making up the remaining portion. The mechanical characteristics and morphology of zinc oxide eugenol are significantly influenced by the ZnO structure [58].

It reveals the beneficial properties of eugenol and its derivatives, the presence of hydrophobic and hydrophilic functional groups within the molecule. Vibrational spectroscopic studies are beneficial in understanding the conformational properties of eugenol [59]. In a study conducted through Raman spectra of eugenol, isoeugenol, and methyl eugenol in the liquid phase, it was shown that although the calculated differences in vibrational spectra for different conformers were small, conformational analysis of these three molecules allowed understanding the biological activities of eugenol and its derivatives [59].

9.2.12 Gutta-Percha

Gutta-percha is the most used core filling material in endodontics. It is a biocompatible, thermoplastic, non-toxic material and is widely used with sealer to strengthen the root canal filling after chemo-mechanical preparation [11]. Its molecular content includes trans-1,4-polyisoprene matrix (ranging from 14.5 to 21.8%), waxes and resins (1-10.4% by weight, added as plasticizing agents), and zinc oxide responsible for its antibacterial properties. (in weighing of 36.6–84.3%), barium sulfate (0–31.2% by weight), along with small percentages of antioxidants and coloring agents [60].

The root canal filling material must maintain its chemical and volumetric stability. However, it may be exposed to chemical conditions during root canal treatment by applying irrigation solutions or different canal pastes [61]. It has been reported that some pastes, especially those containing zinc oxide and eugenol, cause volumetric changes in gutta-percha cones [62]. A published study investigated the effect of high pH on the volume and surface properties of dental gutta-percha cones. The study results showed similar peak distribution of the average Raman spectra of gutta perchas of different groups (control group, pH7, pH12). It was also reported that the areas of the peaks located at 343 and 2850 cm^{-1} were significantly larger in gutta-percha aged at pH12 compared to gutta-percha aged in water. Still, no significant difference was detected between the other peaks evaluated [63].

9.2.13 Resilon

Root canal treatment aims to eliminate bacteria and their residues in the root canal area, prevent recontamination, and ensure a good seal. Recently, methacrylate-based Resilon has been introduced to the market as an alternative to gutta-percha [64]. Degrees of transformation of resin-based materials refer to the monomeric conversion of carbon–carbon double bonds to polymeric carbon–carbon single bonds [65].

The Raman spectroscopic analysis technique was used to determine the chemical composition of Resilon. In a study evaluating the change in Resilon's conversion degree over time, the calculation was made using Raman spectra. While a significant change in the degree of change of Resilon was observed at one week, little change was observed in the subsequent period. The peak height at 1639 cm^{-1} was higher in the uncured Resilon than in the cured one after one week; the peak height at 1609 cm^{-1} was accepted as the internal standard [66].

9.2.14 Mineral Trioxide Aggregate

A suitable and healthy retrograde and orthograde filling material should be non-toxic, biocompatible, insoluble in tissue fluids, and maintain its dimensional stability. Mineral trioxide aggregate (MTA) has two forms, Gray MTA (GMTA) and White MTA (WMTA), with its structure that provides these features. Both formulas basically consist of 75% Portland cement, 20% bismuth oxide, and 5% gypsum (Ca). It contains trace amounts of SiO_2, $K2SO_4$, $Na2SO_4$, and CaO and MgO [67]. It also hardens in water and is a hydrophilic powder that solidifies approximately four hours after settling into dental tissue [11].

MTA cement is a material used in endodontic treatments such as root-end filling, orthograde canal filling, pulpotomy, apexification, apexogenesis, pulp capping, repair of furcation, and root perforations [68].

Calcium and hydroxyl release, water absorption, surface morphology, solubility, and apatite formation abilities of MTAs were investigated [68, 69]. Siboni et al. evaluated two different MTA materials by micro-Raman spectroscopy to test the surface character and calcium phosphate (CaP) nucleation after immersion in HBSS (Hank's Balanced Salt Solution). Micro-Raman analysis of both samples revealed that they contained tricalcium silicate (as alite), dicalcium silicate (belite), and calcium sulfate (as anhydrite) as a fine powder. It also confirmed micro-Raman's ability to nucleate calcium phosphates on their surfaces after immersion in HBSS [68].

9.3 CONCLUSIONS

Examining the structural and chemical features of dental materials with Raman spectroscopy is one of the sensitive techniques. In recent years, there has been a significant improvement in the instrumentation of Raman spectroscopy, leading to increased sensitivity and accuracy based on evolving technological capabilities. At the same time, being non-invasive, it also offers research opportunities in *in vivo* studies. Therefore, it has been used in examining and analyzing dental materials in many studies in the literature. With continued advances in spectroscopic techniques, this technique has potential for further applications.

REFERENCES

1. Kaczmarek K, Leniart A, Lapinska B, Skrzypek S, Lukomska-Szymanska, M (2021). Selected spectroscopic techniques for surface analysis of dental materials: A narrative review. *Materials* 14(10), 2624.
2. Zhang Y and Kelly J R (2017) Dental ceramics for restoration and metal veneering. *Dental Clinics of North America* 61(4),797–819.
3. Meschi N, Patel B, Ruparel N B (2020) Material pulp cells and tissue interactions. *Journal of Endodontics* 46, S150–S160.

4. Revilla-León M, Morillo J A, Att W, Özcan M (2021) Chemical composition, knoop hardness, surface roughness, and adhesion aspects of additively manufactured dental interim materials. *Journal of Prosthodontics* 30(8), 698–705.
5. Ercoli C and Caton J G (2018) Dental prostheses and tooth-related factors. *Journal of Periodontology* 89, S223–S236.
6. Preoteasa E A, Preoteasa, E S, Suciu I, Bartok R N (2018) Atomic and nuclear surface analysis methods for dental materials: A review. *AIMS Materials Science* 5 (4), 781–844.
7. Ramakrishnaiah R, Rehman G Ur Basavarajappa S, Al Khuraif A A, Durgesh B H, Khan A S, Rehman, I Ur. (2015) Applications of Raman Spectroscopy in Dentistry: analysis of tooth structure. *Applied Spectroscopy Reviews* 50(4), 332–350.
8. Raman C, Krishnan K A (1928) New Type of Secondary Radiation. *Nature* 121, 501–502.
9. Otel I (2023) Overall review on recent applications of Raman Spectroscopy technique in Dentistry. *Quantum Beam Science* 7(1), 5.
10. Krafft C, Popp, J (2015) The many facets of Raman spectroscopy for biomedical analysis. *Analytical and Bioanalytical Chemistry*, 407, 699–717.
11. Khan A S, Khalid H, Sarfraz Z, Khan M, Iqbal J, Muhammad N, Fareed M A, Rehman I U (2017) Vibrational spectroscopy of selective dental restorative materials. *Applied Spectroscopy Reviews* 52(6), 507–540,
12. Jones R R, Hooper D C, Zhang L, Wolverson D, Valev V K (2019) Raman Techniques; Fundamentals and Frontiers. *Nanoscale Research Letters* 14, 1–34.
13. Fernández-Galiana Á, Bibikova O, Vilms Pedersen S, Stevens M M (2024) Fundamentals and Applications of Raman-based techniques for the design and development of active biomedical materials. *Advanced Materials* 36(43), e2210807.
14. Do Nascimento G M (2018) Introductory chapter: the multiple applications of Raman spectroscopy. In: Do Nascimento G M (eds). *In Raman Spectroscopy*. InTech p. 1–10
15. Nunes A, Magalhães S (2018) Raman spectroscopy applied to health sciences. In: Do Nascimento G M (eds). *In Raman Spectroscopy*. InTech.
16. Ramírez-Elías M G, González F J (2018) Raman spectroscopy for in vivo medical diagnosis. Raman. In: Do Nascimento G (eds). In *Raman Spectroscopy*. InTech p. 293.
17. Krafft C and Sergo V (2006) Biomedical applications of Raman and infrared spectroscopy to diagnose tissues. *Spectroscopy* 20, 195–218
18. Choo-Smith L P, Hewko M, Sowa M G (2010) Emerging dental applications of Raman spectroscopy. In: Matousek P, Morris M (eds). *Emerging Raman applications and techniques in biomedical and pharmaceutical fields. Biological and Medical Physics, Biomedical Engineering*. Springer, Berlin, Heidelberg p. 263–284.
19. Ko A C T, Choo-Smith L P, Hewko M, Leonardi L, Sowa M G, Dong C C S, Williams P and Cleghorn B (2005) Ex vivo detection and characterization of

early dental caries by optical coherence tomography and Raman spectroscopy. *Journal of Biomedical Optics* 10(3), 031118.
20. Zhang Y, Ren L, Wang Q, Wen Z, Liu C, Ding Y (2022) Raman spectroscopy: A potential diagnostic tool for oral diseases. *Frontiers in Cellular and Infection Microbiology* 12, 775236.
21. Ionita, I (2009) Diagnosis of tooth decay using polarized micro-Raman confocal spectroscopy. *Romanian Reports in Physics* 61, 567–574.
22. Wang Y, Spencer P, Walker M P (2007) Chemical profile of adhesive/caries-affected dentin interfaces using Raman microspectroscopy. *Journal of Biomedical Materials Research Part A* 81(2), 279–86.
23. Zaugg L K, Banu A, Walther A R, Chandrasekaran D, Babb R C, Salzlechner C, Hedegaard M A B, Gentleman E, Sharpe P T (2020) Translation approach for dentine regeneration using GSK-3 antagonists. *Journal of Dental Res*earch 99(5), 544.
24. Vargas-Koudriavtsev T, Durán-Sedó R, Sáenz-Bonilla P, Bonilla-Mora V, Guevara-Bertsch M, Jiménez-Corrales R A, Herrera-Sancho O (2015) Effect of tooth-bleaching agents on phosphate concentration in dental enamel by means of Raman spectroscopy. *Revista Odontológica Mexicana Órgano Oficial de la Facultad de Odontología UNAM* 19(4), 232–239.
25. Khan A S, Hassan K R, Bukhari S F, Wong F S L and Rehman I U (2012) Structural and in vitro adhesion analysis of a novel covalently coupled bioactive composite. *Journal Biomedical Materials Research Part B: Appl Biomaterial* 100 (1), 239–248.
26. Gomes de Araújo-Neto V, Sebold M, Fernandes de Castro E, Feitosa V P, Giannini M (2021) Evaluation of physico-mechanical properties and filler particles characterization of conventional, bulk-fill, and bioactive resin-based composites. *Journal of the Mechanical Behavior of Biomedical Materials* 115, 104288.
27. Döpner S, Müller F, Hildebrandt P, Müller R T (2002) Integration of metallic endoprotheses in dog femur studied by near-infrared Fourier-transform Raman microscopy. *Biomaterials* 23(5), 1337–45.
28. Lubas M (2020) Au interface effecton Ti dental porcelain bond strength investigated by spectroscopic methods and mechanical tests. *Journal of Molecular Structure* 1208, 127870.
29. Lubas M, Przerada I, Zawada A, Jasinski J J, Jelen P (2020) Spectroscopic and microstructural investigation of novel Ti–10Zr–45S5 bioglass composite for dental applications. *Journal of Molecular Structure* 1221, 128545.
30. Spies B C, Zhang F, Wesemann C, Li M, Rosentritt M (2020) Reliability and aging behavior of three different zirconia grades used for monolithic four-unit fixed dental prostheses. *Dental Materials* 36(11), e329–e339.
31. Ubaldini A L M, Pascotto R C, Sato F, Soares V O, Baesso M L (2020) Mechanical and chemical changes in the adhesive-dentin interface after remineralization. *The Journal of Adhesive Dentistry* 22(3), 297–309.
32. Par M, Spanovic N, Mohn D, Attin T, Tauböck T T, Tarle Z (2020) Curing potential of experimental resin composites filled with bioactive glass: A comparison between Bis-EMA and UDMA based resin systems. *Dental Materials* 36(6), 711–723.

33. Lopes C B, Pinheiro A L, Sathaiah S, Duarte J, Cristinamartins M (2005) Infrared laser light reduces loading time of dental implants: a Raman spectroscopic study. *Photomedicine and Laser Surgery* 23(1), 27–31.
34. Lach S, Jurczak P, Karska N, Kubiś A, Szymańska A, Rodziewicz-Motowidło S (2020) Spectroscopic methods used in implant material studies. *Molecules* 25(3), 579.
35. Lupi S M, Galinetto P, Albini B, Di Ronza E, Rizzo S, Rodriguez y Baena R (2020) Micro-Raman spectroscopy of dental implants subjected to different surface treatments. *Applied Sciences* 10(7), 2417.
36. Malhotra R, Han Y M, Morin J L P, Luong-Van E K, Chew R J J, Castro Neto A H, Nijhuis C A, Rosa V (2020) Inhibiting corrosion of biomedical-grade Ti-6Al-4V alloys with graphene nanocoating. *Journal of Dental Research* 99(3), 285–292.
37. Surana P, Singh Dhull K, Arya A, Samreen S, Rajan M, Parihar A S (2024) Bio-ceramics application in Dentistry. *Bioinformation* 20(2), 136–139.
38. de Carvalho R F, Cruz F L G, Antunes D P, Júnior E G T, de Oliveira L F C, de Paulo M V Q, Devito K L (2018). In vitro analysis of dental ceramics: evaluation of the radiopacity and chemical composition by Raman spectroscopy. *Brazilian Dental Science* 21(2), 164–171.
39. Okulus Z, Buchwald T, Szybowicz M, Voelkel A (2014) Study of a new resin-based composites containing hydroxyapatite filler using Raman and infrared spectroscopy. *Materials Chemistry and Physics* 145(3), 304–312.
40. Kırmalı Ö (2014) Dental ceramics used in dentistry. *Cumhuriyet Dental Journal* 17(3), 316–324.
41. Jestel N L, Morris M D, O'Brien W J (1998) Depth-resolved Raman microprobe examination of a commercial dental porcelain exposed to a simulated oral environment. *Dental Materials* 14(5), 375–82.
42. Wulfman C, Sadoun, M, De La Chapelle M L (2010) Interest of Raman spectroscopy for the study of dental material: The zirconia material example. *IRBM* 31(5–6), 257–262.
43. Munoz Tabares J A, Anglada M J (2010) Quantitative analysis of monoclinic phase in 3Y-TZP by Raman spectroscopy. *Journal of the American Ceramic Society* 93(6), 1790–1795.
44. Presser V, Keuper M, Berthold C, Nickel K G (2009) Experimental determination of the Raman sampling depth in zirconia ceramics. *Applied Spectroscopy* 63(11), 1288–1292.
45. Gatin E, Iordache S M, Matei E, Luculescu C R, Iordache A M, Grigorescu C E A, Ilici R R (2022) Raman spectroscopy as spectral tool for assessing the degree of conversion after curing of two resin-based materials used in Restorative Dentistry. *Diagnostics* 12(8),1993.
46. Iordache S M, Iordache A M, Gatin D I, Grigorescu C E A, Ilici R R, Luculescu C R, Gatin E (2024) Performance assessment of three similar dental restorative composite materials via Raman spectroscopy supported by complementary methods such as hardness and density measurements. *Polymers* 16(4),466
47. Rocca G T, Rizcalla N, Krejci I (2013) Fiber-reinforced resin coating for endocrown preparations: A technical report. *Operative Dentistry* 38(3), 242–248.

48. Zhang M and Matinlinna J P (2012) E-glass fiber reinforced composites in dental applications. *Silicon* 4 (1), 73–78.
49. Bonon A J, Weck M, Bonfante E A, Coelho P G (2016) Physicochemical characterization of three fiber-reinforced epoxide-based composites for dental applications. *Materials Science and Engineering C: Materials for Biological Applications* 69,905–13.
50. Ozer F, Blatz M B (2013) Self-etch and etch-and-rinse adhesive systems in clinical dentistry. *Compendium of Continuing Education in Dentistry* 34(1), 12–4.
51. Santini A, Miletic V, Koutsaki D (2012) Degree of conversion of three fissure sealants cured by different light curing units using micro-Raman spectroscopy. *Journal of Dental Sciences*, 7(1), 26–32.
52. Navarra C O, Cadenaro M, Armstrong S R, Jessop J, Antoniolli F, Sergo,V., Di Lenarda R and Breschi L (2009) Degree of conversion of filtek silorane adhesive system and clearfil SE bond within the hybrid and adhesive layer: An in situ Raman analysis. *Dental Materials* 25 9), 1178–1185.
53. Miyazaki M, Onose H, Moore B K (2002) Analysis of the dentin–resin interface by use of laser Raman spectroscopy. *Dental Materials* 18(8), 576–580.
54. Young A M, Sherpa A, Pearson G, Schottlander B, Waters D N (2000) Use of Raman spectroscopy in the characterisation of the acid-base reaction in glass-ionomer cements. *Biomaterials* 21(19), 1971–9.
55. Gryksa M, Eriksson M, De Basso N, Shen Z (2020). Phosphate dental cements aged in vivo up to 25 years. *Advances in Applied Ceramics* 119(5–6), 338–347.
56. Frost R L (2004) An infrared and Raman spectroscopic study of natural zinc phosphates. *Spectrochimica Acta Part A: Molecular and Biomolecular Spectroscopy* 60(7), 1439–1445.
57. Paluszkiewicz C, Kwiatek W M, Dugon E, Weseucha-Birczynska A, Piccinini M (2009) Surface study of selected biomaterials using vibrational spectroscopy *Acta Physica Polonica A* 115 (2), 533–536.
58. Bakhori S K M, Mahmud S, Mohamad D, Masudi S M, Seeni A (2019) Surface morphological and mechanical properties of zinc oxide eugenol using different types of ZnO nanopowder. *Materials Science and Engineering C: Materials for Biological Applications* 100, 645–654.
59. Chowdhry B Z, Ryall J P, Dines T J, Mendham A P (2015). Infrared and Raman spectroscopy of eugenol, isoeugenol, and methyl eugenol: Conformational analysis and vibrational assignments from density functional theory calculations of the anharmonic fundamentals. *The Journal of Physical Chemistry A*, 119(46), 11280–11292.
60. Taddei P, Di Foggia M, Zamparini F, Prati C, Gandolfi M G (2023) Guttapercha improves in vitro bioactivity and dentin remineralization ability of a bioglass containing polydimethylsiloxane-based root canal sealer. *Molecules* 28(20), 7088.
61. Tagger M, Greenberg B, Sela G (2003) Interaction between sealers and gutta-percha cones. *J Endod* 29, 835–7.

62. Michaud R A, Burgess J, Barfield R D, Cakir D, McNeal S F, Eleazer P D (2008) Volumetric expansion of gutta-percha in contact with eugenol. *Journal of Endodontics* 34(12), 1528–32.
63. Atmeh A R, AlMadi M, AlShwaimi E (2022) Alkaline conditions can affect the volume and chemical characteristics of dental gutta-percha. *Australian Endodontic Journal* 48(1),144–150.
64. Shipper G, Ørstavik D, Teixeira F B, Trope M (2004) An evaluation of microbial leakage in roots filled with a thermoplastic synthetic polymer-based root canal filling material (Resilon). *Journal of Endodontics* 30(5), 342–7.
65. Ferrancane J L (1985) Correlation between hardness and degree of conversion during the setting reaction of unfilled dental restorative resins. *Dental Materials* 1(1), 11–4.
66. Wu W C, Shrestha D, Wei X, Ling J Q, Zhang W H, Chen J (2010). Degree of conversion of a methacrylate-based endodontic sealer: a micro-Raman spectroscopic study. *Journal of Endodontics*, 36(2), 329–333.
67. Malhotra N, Agarwal A, Mala K (2013) Mineral trioxide aggregate: a review of physical properties. *Compendium of Continuing Education in Dentistry* 34(2), e25–32.
68. Siboni F, Taddei P, Prati C, Gandolfi M G (2017) Properties of NeoMTA Plus and MTA Plus cements for endodontics. *International Endodontic Journal* 50(Suppl 2), e83–e94.
69. Gandolfi M G, Siboni F, Primus C M, Prati C (2014) Ion release, porosity, solubility, and bioactivity of MTA Plus tricalcium silicate. *Journal of Endodontics* 40(10), 1632–7.

CHAPTER 10

Examining the Microstructural and Nanostructural Properties of Dental Materials Using Raman Spectroscopy

Approaches, Uses, and Future Directions

Preeti Pandey

10.1 INTRODUCTION

Dental enamel, the strongest and most mineralized tissue in the human body, covers and shields the tooth's anatomic crown.[1–3] It protects the pulp, dentin, and inner layers from the stresses applied while chewing. Enamel mineralization, an important characteristic, strongly correlates with the frictional behavior of tissues, including bone and teeth.[4–6] About 1–3% of dentin is water, 95–96% is a mineral in the form of hydroxyapatite (HAp), and the last 1–2% is an organic material, mostly protein in the form of tufted in, enameling, amelogenin, and ameloblastin.[1–3,6–9] Researchers are considering some vibrational spectroscopy techniques, such as Fourier

DOI: 10.1201/9781003506805-10

transform infrared (FTIR), near infrared, and Raman as potential substitutes for more established methods of diagnosing biological tissues.[1] Since FTIR spectroscopy detects changes in a molecule's dipole moment and Raman spectroscopy analyses changes in a molecule's polarization, the two techniques are complementary to one another. Recent advancements in apparatus and fiber optic probe technology have led to increased use of Raman spectroscopy in dentistry and other biological sciences.[2–5] Its applications in biomaterials, structural chemistry, and surface characterization have also increased.[6] Here are some of the benefits of Raman spectroscopy: It is non-invasive, non-destructive, yields consistent results, and requires little to no sample preparation.[1,6] Researchers have conducted extensive research to define and detect disorders of rigid tissues, including tooth tissues, as well as healthy and sick bones, thereby expanding their usefulness beyond soft tissues.[7] The biomedical field initially introduced Raman spectroscopy, despite encountering numerous challenges. The fluorescence of biological tissues and the scarcity of sensitive equipment were the main challenges in Raman spectroscopy (RS) faced when it first entered the biomedical field. Over the past ten years, the advancements in RS have attracted significantly more favourable attention in the biomedical field than advancements in infrared spectroscopy. This is because, in contrast to IR spectroscopy, where there is a huge spectral band,[6] water exhibits less interference in Raman spectra. Biological soft and rigid tissues both contain substantial amounts of water. Raman spectroscopic imaging has been used in various ways to characterize areas of interest in the medical and dental fields. Finding the precise peaks in the spectrum that serve as fingerprints to differentiate between healthy and ill tissues is the ultimate objective of all these alterations. This article will discuss how to diagnose hard and soft tissue problems in teeth using RS in dentistry. It will also highlight important peaks and the related spectrum assignments for those peaks. This work can serve as a reference for future investigations to determine pertinent RS peaks for the diagnosis of sick oral muscles and soft tissues. Raman spectra can be used for analysis of any crystalline structure, particularly a human tooth, to determine its chemical makeup. It is also a very selective method of studying molecular species. It only analyses the dentinal region of the tooth, and an individual's chemical makeup changes with age. Fluorescence spectra from organic materials frequently outweigh the relatively weaker Raman signals in RS. Only enamel has been the subject of Raman spectroscopy research due to its extremely low organic content.

10.2 FUNDAMENTALS OF RAMAN SPECTROSCOPY

Scientists have shown that Raman spectroscopic method, a recently created bio-photonic device, is a very useful molecular spectroscopic analysis tool with numerous applications in various research domains. Raman spectroscopy is still a relatively young analytical technique, despite Dr. C. V. Raman receiving the 1928 Nobel Prize for discovering the Raman scattering phenomenon. It is widely used in numerous industries, including the medical sector. Vibrational spectroscopy is currently considered a valuable optical technology for biological applications because it can safely analyze molecular composition and consistently and uniquely detect fingerprints. What makes this technique so special is the way a small amount of light scatters in different ways when it reaches the sample under study. When these light photons scatter off molecules or crystals, they spread elastically. Most scattered photons have wavelengths and frequencies that match those of the incoming photons. However, a small fraction of light—roughly 1 in 107 photons—scatters at optical frequencies that are different from, and frequently lower than, the frequency of the incident photon. The Raman effect causes the inelastic scatter. Photons disperse elastically through a process called Rayleigh scattering. Alternatively, we term it Raman scattering if the photons diffuse inelastically. When the scattered photon frequency is lower than the incident photon frequency, Stokes scattering takes place. Anti-Stokes scattering is the mechanism that occurs when the photons scatter at a higher frequency, as seen in Figure 10.1.[8] It is a handy tool for studying inorganic surface tissueslike carbonate and phosphate because it can provide exact spectroscopic information on minerals by detecting the various frequencies of their vibrational modes.[10]

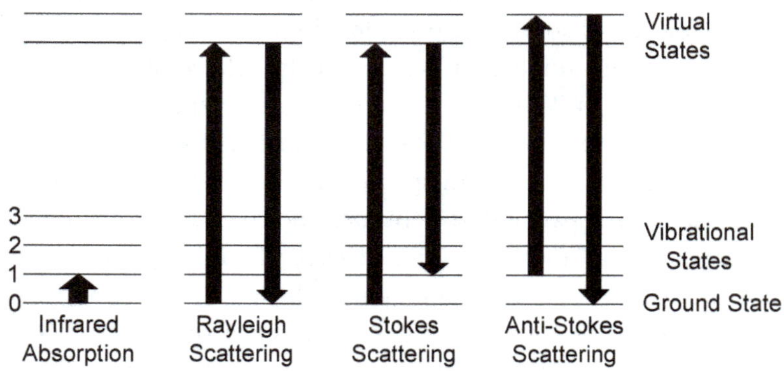

FIGURE 10.1 Energy-level diagram for Raman scattering: (a) Rayleigh, (b) Stokes, and (c) anti-Stokes scattering.[8]

Biomedical research is increasingly relying on RS due to its ability to operate in the near-infrared (NIR) range, its ability to selectively detect specific biochemicals, its non-intrusive technique, its compatibility with water, and its remote or in-person use via fiber optics.[9,10] This versatile, often applied method simplifies the evaluation of tissue samples in any state—dry or in solution—and performs well in physiological contexts. It can conduct tests in vivo, depending on the circumstances, and is highly sensitive to morphological changes such as structural and chemical alterations, even though it is not sensitive to the form, state, or test material. Optical fibers can effectively transfer and gather laser signals and Raman-dispersed photons over short or long distances for local analysis by using the laser beam as a monochromatic light source. Within minutes, the Raman spectrum may identify and quickly identify various molecules based on their chemical composition and structural stability arrangements, as well as minerals based on their morphology and crystal orientation, depending on the region under inquiry. You can get Raman spectra from materials that are less than 1 μm in diameter to look at surface microstructures and orientation, as well as micron-level pollutants like porphyrins that are caused by flaws, microbes, or very thin coatings (like dental films). Moreover, users can finish automated chemical analysis rapidly, regardless of their level of experience (starting, advanced, or pro).[9–18] During a Raman spectroscopy investigation, a laser beam typically focusses on the sample to record the energy profile of the scattered light (Stokes scattering). In more complicated or mixed materials, such as solutions or combinations of many compounds, we were able to get spectra that comprised the Raman energy of every component. By analyzing the resulting band intensities, we were able to detect the same amounts and monitor changes in structural or biochemical composition.[16–18] Whether a sample is a compound or a chemical element, activating the vibrational modes of one or more different molecules yields a specific range for each sample under investigation. A fingerprint zone, comprising bands of interest and zones at both high and low frequencies, composes each Raman spectrum.

10.3 EXAMINING THE MICRO- AND NANOSTRUCTURE PROPERTIES OF A DENTAL IMPLANT

Dental material architectures are robust and rigid, making it difficult to cross-section these materials mechanically. However, we can generate ultra-thin, homogeneous lamella sections using focused ion beam-scanning electron microscope (FIB-SEM) dual-beam devices. This work employed complementary imaging and diffraction analysis, along with transmission electron microscopy (TEM) and FIB-based advanced characterization techniques, to reveal various morphological properties of tooth tissue. Scanning electron microscopy-energy dispersive X-ray spectroscopy (SEM-EDS) and Raman scattering methods also clarified the component distribution and changes in the chemical structure of the dental tissues. At the point where the dentin and enamel layers meet, electron microscopy showed that the enamel was denser and had more crystals, while the dentin layer was porous, fibrillar, and didn't have any long-range order. Specifically, there were hardly any variations in the elemental distribution found in the EDS mapping and line scan studies. Raman analysis revealed that the chemical makeup of both tissues was similar, except for dentin, whose tubular structure and organic components caused spectral background effects in the spectrum. To investigate the surface properties of dental materials, researchers widely use spectroscopic techniques such as FTIR spectroscopy, Raman spectroscopy, ultraviolet radiation, visible spectroscopy (UV-Vis), X-ray spectroscopy, and mass spectrometry (MS). Here are further examples of studies on materials for tooth fillings, procedures for preparing cavities, surface preparation for dental prostheses, and fractographic investigations of dental implants. The mentioned studies show that the aforementioned methods can serve as valuable resources to enhance the research potential of dental materials. With the development of nanotechnology, adding nanoscale fillers to dental resins appears to be a potential way to enhance therapeutic qualities and give dental materials better physicochemical characteristics. Nanotubes are increasingly gaining use due to their superior mechanical qualities, their ability to transport and administer medications, and their bioactive qualities. These characteristics are dependent on the use and makeup of nanotubes. This scoping review aims to summarize earlier research on the use of nanotubes in materials based on restorative resins. Here, our primary objectives are to[19] identify the most commonly used nanotubes in the creation of these dental materials,[20] to confirm the molecules or particles associated with these nanotubes, and[21] to report on the main outcomes and goals of the nanotube incorporation process.

10.3.1 Enamel

Two tough tissues that make up human teeth, namely the enamel and dentin layers, differ structurally even though their chemical makeup is extremely similar. For decades, researchers have regularly used X-ray tomography analyses, which can only provide spatial resolution down to hundreds of nanometers, to analyze tooth tissues in detail. However, electron microscopy-based characterization techniques allow for chemical analysis and nanoscale spatial resolution imaging of samples. FIB microscopy specifically uses a focused, direct beam of high-energy ions to accomplish site-specific micro- and nanoscale machining and characterization. This method is useful for characterizing ultra-hard materials that are difficult to work with mechanically. TEM and SEM can analyze an extensive range of materials elementally and with great resolution.[22] Because dental tissues are very hard and stiff, they can be processed with an ion beam. This is especially true when studying the underlying structural morphology and chemistry of the individual layers and their interfaces. Enamel, a composite material consisting of both an organic and a mineral phase, is the hardest biological substance found in the human body. Ninety-six to ninety-six weight percent of the mineral phase is composed of large hexagonal hydroxyapatite crystals that are magnesium phosphate salts. Similarly, rod-shaped structures with a cross-sectional diameter of 3–6 μm are known as "enamel prisms," produced by orientated crystals. Water contributes the remaining three weight percent of the enamel, with protein and the organic matrix accounting for around one weight percent.[23]

10.3.2 Dentin

Since dentin makes up the majority of the tissue in human teeth, changes to its structure are likely to cause the tooth to become more fragile.[24] Dentin is a biological composite consisting of three phases: fluid, type I collagen fibrils (about 30% by volume), and mineral phase (hydroxyapatite, approximately 50% by volume).[25] In recent years, a significant amount of research has been conducted using cutting-edge techniques to examine the morphology of dentin and other calcified tissues.[26–28] X-ray diffraction, narrow-angle X-ray scattering, atomic force microscopy, transmission electron microscopy, and other advanced techniques for characterization were used to carefully examine dentin's size, shape, orientation, molecular and chemical composition, and mechanical properties.[23,24,29–33] The micro- and nano-structural changes brought on by aging, caries, and sclerosis, as well as the results of dental operations, can significantly affect how strong teeth are. To forecast how microstructural alteration may affect tooth strength, novel approaches have recently analyzed several altered forms of dentin.[34] However, there has been no attempt to investigate any changes in the form, composition, or size of the noncrystalline apatite mineral within the devitalized dentin. In this study, we used atomic force microscopy, scanning electron microscopy/energy dispersive X-ray analysis, and micro-Raman spectroscopy to investigate the nano-structural and compositional causes of the mechanical breakdown of teeth that had lost their vitality.

10.3.3 Restorative Materials

What we now call ceramic comes from the Greek word keramos, which means "a potter" or "pottery." Keramos comes from an older Sanskrit word that means "to burn." In 1967, Gilman[35] described pottery as an earthy substance, typically composed of silicates. It can be defined as a combination of a nonmetallic element (usually oxygen) with at least two metals. Ceramics are inorganic, nonmetallic, and generally crystalline solids, according to the American Ceramic Society. Examples of metallic and nonmetallic elements that combine to make ceramics are oxygen and aluminum, calcium and oxygen (calcium oxide or CaO), and nitrogen and silicon. Biocompatible materials have been used in the creation of ceramic dental restorations. The interest in all-ceramic restorations has increased dramatically during the last ten years. This is because materials are becoming stronger, more resilient, and more aesthetically pleasing thanks to new processing technologies like hot pressing and CAD/CAM (computer-aided design/computer-assisted manufacturing).[36] Dentistry frequently employs ceramics due to their natural appearance and exceptional mechanical, optical, thermal, and chemical properties.[37] Dental ceramics have many excellent qualities, such as being biocompatible, chemically neutral, aesthetically pleasing, strong, not letting plaque build up, insulating, resistant to wear, and maintaining their color.[38,39] However, oral all-ceramic restorations' fragility, simple damage, low tensile strength, and durable breakability limit their use in many situations.[40] Many flaws on the outside and inside of ceramics can happen because of temperature, chemical, or mechanical processes. These flaws can concentrate stress, making ceramics less strong. These stresses can cause cracks at the flaw site to appear and spread, causing the structure to fail catastrophically.[41] In recent years, researchers have experimented with new materials and methods to address these issues. This review talks about how dental ceramics have changed over time. We explore the background, the method of production, the classification, and the mechanical and physical characteristics of the material. The main focus is on how these novel materials and methods of operation are resolving the above-mentioned issues. It covers data on longevity, clinical research, and preparation. It discusses how all-ceramic restorations have evolved, with a focus on high-strength core ceramics and resin-bonded glass-ceramic restorations. Over time, people have become more interested in using ceramic materials in dentistry. This is because ceramics have special qualities, and patients want better-looking and metal-free options. In the field of dental science, ceramics are nonmetallic artificial structures primarily composed of oxygen compounds containing one or more metallic or semi-metallic elements, such as silicon, sodium, zirconium, titanium, and aluminum. Ceramics includes a huge range of materials, making it difficult to give a short description. During the early stages of implant dentistry, experts used ceramics to enhance osseointegration by enclosing endosseous metal implants. For the past ten years, dental ceramics have employed high-tech methods, resulting in the development of new materials such as heat-pressed, injection-molded, slip-cast, and glass ceramics. This book chapter's goal is to look at how new materials and methods have helped make all-ceramic dentistry restorations better. The chapter discusses significant changes in the materials used for all ceramic repairs. Some of these improvements include better handling methods and stronger materials themselves. We look at the main crystalline forms of some new ceramics, such as sanidine, spinel, hydroxyapatite, leucite, alumina, forsterite, zirconium oxide, and vermiculite. Lastly, we provide an extensive collection of data on known flexural strengths for different types of ceramic goods. Over the last 15 years, tooth implants have utilized multiple types of ceramic coatings[42] (Table 10.1). We used both bioactive ceramics, such as calcium phosphates and bioglasses, and inactive ceramics, such as aluminium oxide and zirconium oxide. Coatings with thicknesses ranging

TABLE 10.1 Ceramic Materials Available As Dental Implants and Coatings

Materials	Chemical Composition
Hydroxylapatite (HA)	$Ca_{10}(PO4)_6(OH)_2$
Tricalcium phosphate (TCP)	$\alpha,\beta,Ca_3(PO_4)_2$
Bioglasses	SiO_2-CaO-Na_2O-P_20_5-MgO, *etc.*
Aluminium oxide	Al_2O_3
Zirconium oxide	ZrO_2

Source: Adopted from Lacefield, 1998[42]

from 1 to 100 μm can be produced thinly or densely using various processes. Metal devices can be coated using a variety of methods, including plasma spraying, biomimetic precipitation, sputter deposition, sol-gel coating, and electrophoretic deposition.[43] For individuals without teeth, full-arch, partial-tooth, or single-tooth dental restorations can be stabilized with ceramic or polymer/ceramic composite crowns put on dental implants.[44] We selected these materials because of their distinct composition, production technique, and greater quality when compared to traditional dental materials. The unique properties of the material allow for a wide range of applications in regenerative medicine and orthodontic implantology.

Mercury-based dental amalgam restorations have a lengthy history. China is believed to have used the earliest tin and mercury dental restorations circa 600 AD. France introduced silver-mercury restorations to the West for the first time in the 1830s.[45] Dental professionals in the United States who employed amalgam were subject to malpractice claims from those who did not in the 1850s. This time frame was known as the "Amalgam Wars." The dental community did not instantly embrace Dr. G.V. Black's comprehensive 1896 scientific article endorsing amalgam use.[46] Particularly in the last 30 years, there has been criticism of the use of mercury in dental silver amalgam restorations and the subsequent mercury leakage from these restorations.[47] For the past 150 years, dental amalgam has been one of the most popular restorative materials. Due to its numerous benefits, dental amalgam remains a widely used substance worldwide, especially in situations where caries risk is higher. Its outstanding mechanical qualities (malleability, strength, and ease of application) and durability are a few of the most notable ones.[48–50] Therefore, experts especially recommend it for regions such as the posterior areas, large carious cavities, and individuals with poor dental hygiene, which are subject to strong mechanical loads and do not require a high level of aesthetics.[51] Additional benefits mentioned include affordability, simplicity of use, durability, reduced susceptibility to clinical procedures, and bacteriostatic action.[52–55] The two main components of dentistry's amalgam are alloy powder, which makes up 50% of the composition and contains smaller amounts of metallic silver, tin, copper, and other metals; less than 3%; and mercury.[56,57] Several times, amalgam's usage in dentistry has generated controversy due to its high mercury content, ugly look, and environmental contamination from poor disposal and preservation.[58] The potential health effects of amalgam fillings and mercury exposure have been the subject of much discussion. Dentists are most likely to come into contact with mercury, yet they usually don't appear to care too much about it or the potential harm it could bring to the environment. Many therapeutic contexts prefer amorphous cement over direct restorative materials like glass ionomers and composites because of its mechanical and physical properties, ease of application, and relative affordability. One of its primary constituents, mercury, poses a serious risk to both people and the environment. The four adverse effects of dental amalgam are toxicology, allergic responses, environmental problems, and mouth galvanism.[59] Discusses the environmental effects of dental amalgam, specifically focusing on the mercury content of the material.

10.3.4 Nanofillers in Dental Composite

A new type of composite resin, known as "Nanofillers composites," has entered the dentistry market due to recent advancements in nanotechnology. Researchers first reported composites using nanofillers in the early 1990s.[60] A new dental restorative composite became available, using anodic porous alumina nanostructured materials as its foundation. While its bulk characteristics, such as its ability to transport drugs and reduce the effects of aging, are encouraging, nothing is known about this material's surface characteristics. Nevertheless, there is still a need for improvement in several areas regarding dental composites, including mechanical strength, biocompatibility, and surface characteristics.[61,62] Surface properties are a major focus of current scientific research in the field of dental composites because they have a direct impact on the restoration's durability and performance over the short and long terms.[63] This is due to the direct relationship between their success and the restoration. It is especially good for the mouth because the rough surface of restorative composites encourages bacteria to stick to it, salivary proteins to stick to it, and bacterial plaque to form, all of which cause infiltration and secondary cavities.[64,65] It is difficult to brush teeth on a very big, uneven, and undulating surface while still keeping the area reasonably clean.[66] It's also critical to remember that the roughness of dental composites is linked to a reduction in the repair's aesthetic appeal. The significance of characterizing these surfaces at the nanoscale has been highlighted in several publications.[67–69] Moreover, the adhesion strength is ultimately increased by the surface roughness's significant impact on the integration and bonding of the restorative composite to the oral tissue cavity. Composite paste fillers can now contain more filler because nanosized inorganic fillers reduce the amount of organic resin matrix while maintaining outstanding clinical handling qualities. The field of nano-filled materials has seen a notable improvement in recent times. Considerable progress has been made when composite resin was mixed with nanofillers. According to Rezvani et al.[70] the nanosized fillers can be either separate particles with sizes between 5 and 100 nm or groups of primary nanoparticles that are fused, with cluster sizes much larger than 100 nm. The right surface treatment can control high quantities of nanoparticle clusters. This filler enhancement increases the surface area and thickens the consistency of the composite resin paste, resulting in smaller filler particles and lower loading levels.[71] After mounting the samples on glass slides and applying sputter coating, we used X-ray diffractometry to analyze them. Each of them used an XRD 3003 PTS diffractometer with Cu Ka radiation (λ = 1.5418A°. SEM stands for scanning electron microscopy. We examined the surface appearance of the materials using a variety of microscope magnifications.

10.3.4.1 Graphene Nanofiller

With the help of SEM, the rough surface of the GRP film showed an edge plane of the rolled-up and wrinkly graphene sheet structure[72] (Figure 10.2).

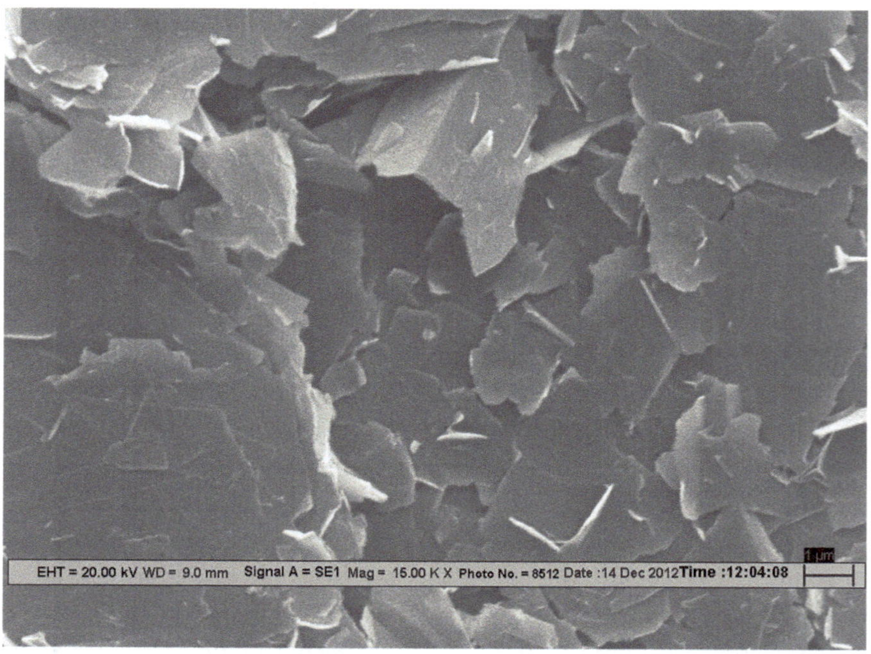

FIGURE 10.2 2-Sem Image of Graphene Film at 160,00 KX.

It is commonly known that functional sites like epoxy and hydroxyl groups can be found in the edge plane of crumbled graphene sheets. Phase characterization is an X-ray diffraction (XRD) pattern at 2θ ranges between 10° and 70°. The XRD pattern of GRP (Figure 10.3) shows a prominent diffraction peak at 2θ = 27°, indicating that the particles are crystalline.

10.3.4.2 Mesoporous Zirconia Fillers

Zirconia is well known in the ceramics sector for its exceptional hardness and resistance to common fractures. Its sub-micron grain size also makes it possible to preserve the sharpness of the edge and produce superb surface finishes. As electrical technology developed at the same time as zirconia ceramics, dental science and industry were able to make progress. Zirconia's biocompatibility and aesthetic qualities have made it increasingly desirable in biomedical applications, particularly surgical implants. Due to the advantageous physical, esthetic, biological, and corrosion properties of zirconia, its use in dentistry and medicine has increased dramatically over the past ten years. However, Zirconia typically has surface areas of 100 m²/g or less, which is not very large compared to conventional catalysts and supports made of silica or alumina. Amorphous zirconia can achieve higher surface areas (200–300 m²/g) but at the cost of decreased heat stability. The thermal stability and large surface area of ZrO_2 make it an appealing option for heterogeneous catalytic applications. The SEM picture of ZrO_2 is shown in Figure 10.4, which also demonstrates its structure

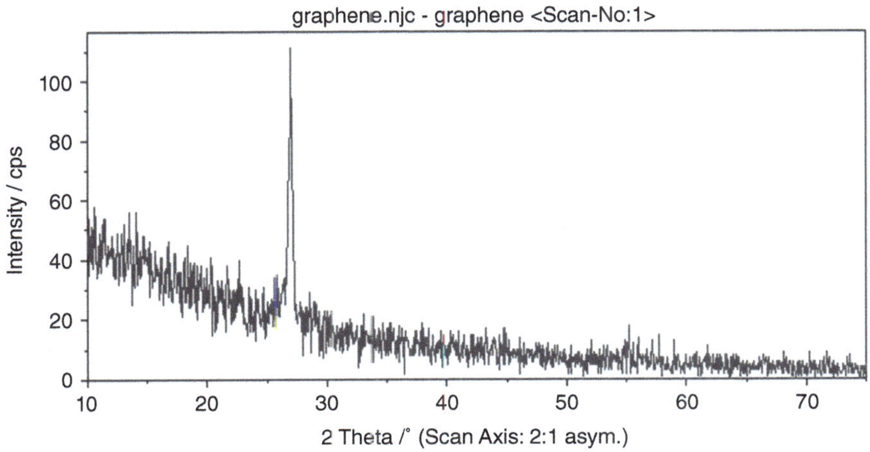

FIGURE 10.3 XRD analysis of GRP.[72] The samples were prepared as thin films on glass slides by drop-casting and room-temperature drying.

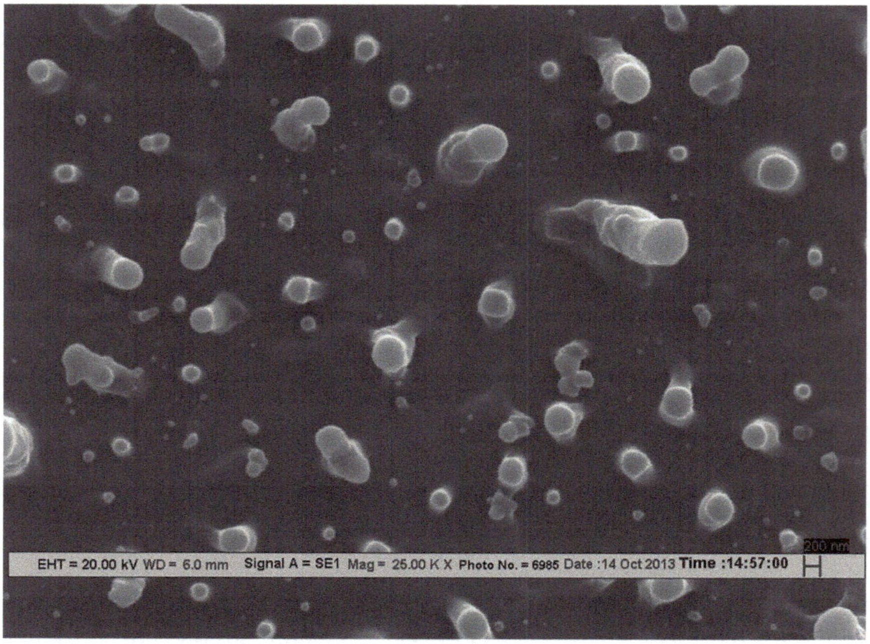

FIGURE 10.4 SEM image of zirconium oxide nanoparticles composite.

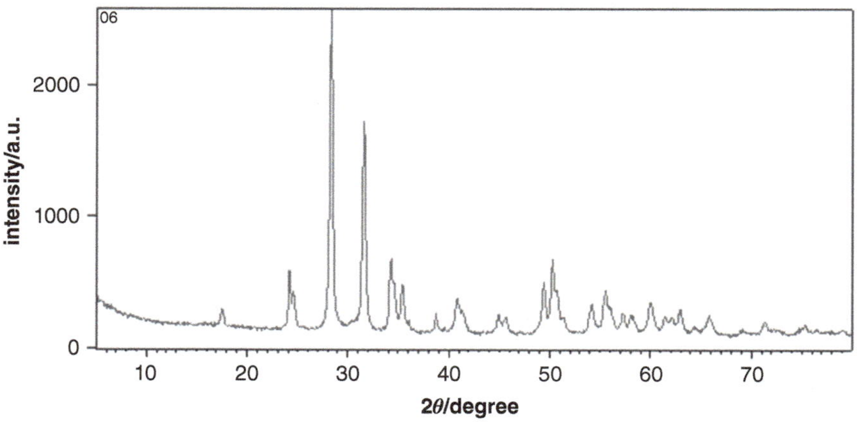

FIGURE 10.5 XRD pattern of ZrO_2 for dental applications.[73]

and morphology. The uneven forms and ovulation were the predominant features of the zirconia's morphology. Figure 10.5 displays the zirconia nanoparticles' XRD pattern. Two ranges of 10°–70° absorb the diffraction peaks. Zircon is crystalline, as shown by the two clear wide peaks in the X-ray diffraction patterns of pure ZrO_2 nanoparticles at scattering angles of 28.29° and 31.57°, which are equal to 3.1 Å and 2.8 Å between the planes of the particles.[73]

10.3.5 Resin-Based Composites

When treating children's teeth, composite materials with a plastic foundation are necessary. Buonocore[74] was the one who initially suggested acid-etching. It maintains attractive dental fillings in primary and permanent teeth in place. We can now make resin materials that not only appear better but also wear better, hold colors better, and enable repairs to stay intact for the required amount of time, all thanks to breakthroughs in resin technology. Since the 1990s, composite materials, or RBCs, have grown in popularity. Because RBC may mirror the structure of teeth, it may be advantageous to both patients and dentists. Use resin-based techniques for larger preparations to minimize amalgamation and RBC quality variation. Adjusting the size, filling the padding, and adding silane resolved the initial wear issues in the priority RBC repairs. RBC's diverse viscosities, dispersion methods, polymerization procedures, and ion leachability allow it to offer a wide range of filler particle sizes, colors, and opacities. Understanding the polymerization process of methacrylate-based dental restorative materials is becoming more and more crucial. This has led to the development of numerous novel non-traditional curing sources and techniques, including lasers and plasma arc lamps, which assist in reducing polymerization stress. Furthermore, it is claimed that resin-based composites (RBCs) can cure at thicknesses far greater than those advised by industry standards. The degree of polymerization in each of these situations is relevant to examine since it influences multiple parameters, such as wear,[75] fracture toughness,[76] toughness, polymerization solubility,[77] and ultimately mechanical and fluid mechanical characteristics,[78] and more. RBC fillings are beneficial for two major reasons: they blend in with your teeth's color and have a fantastic appearance.[79] We provide a range of RBC translucency hues to enhance the appearance of the restoration and make it easier to match the tint to neighboring teeth. Numerous investigations on the properties of epoxy composite dental restorations have demonstrated to dentists their potential for many years to come.[80] Improvements in the manufacturing process for resin composites have improved clinical outcomes.[81, 82] The use of resin composite directly on the chair side for cavity treatment in severely stressed posterior teeth has significantly increased recently.[83] Gold or ceramic fillings require more surgery and are more expensive than plastic fillings. Thanks to technological advancements, resin mixes can now cling strongly to the hard tissues of teeth. However, resin composite restorations have not been shown to be successful in highly stressed parts of the mouth. The primary reasons for this are mechanical problems, a higher chance of breaking, and shrinkage that occurs during polymerization. Improvements in the production process of resin composites have led to better clinical outcomes.[84]

10.3.6 Environmental Effects

Their primary attributes are the mechanical qualities, biocompatibility, corrosion resistance, and ease of workability of dental alloys. These alloy properties are the result of a combination of many metals.[85,86] Amalgam fillings are still the most widely used materials for dental restorations, despite significant doubts regarding their biological effects and a heated debate on their use and hazards. Despite the presence of heavy, unwanted metals such as mercury and numerous other metals with significant biological toxicity, dental fillings frequently use amalgams due to the lack of practical substitute materials and their long-standing success.[87] We can use chemical and electrochemical test methods to analyze the microstructure and chemical composition of alloys, which significantly influence their corrosive behavior.[88] Since Darwell[85] showed that dental amalgams can corrode in concentration cells using electrochemical methods, it has become more important to study the corrosion potential of amalgams both in the lab and in living organisms.[89] According to several studies, dental alloys must be able to tolerate variations in temperature (0–65 °C), moisture, and pH (2–11) caused by food eating and oral environment breaks. It is contingent upon their makeup, potential values, strain, degree of oxidation, roughness of the surface, pH values, intermediate temperature, solvent mixing velocity, and inhibitor presence.[90–94] An appropriate root-end filling material should not be affected by tissue fluid, which in an infected area may be acidic. In addition, it should be dimensionally stable, biocompatible, stick to the root-end cavity walls, resist dislocating stresses, and stop bacteria from passing through.[95] Most of these essential qualities are present in mineral trioxide aggregate (MTA), the suggested root-end filling material.[96] Treating immature teeth with necrotic pulps as an apical barrier is also advised. Changes in the pH of host tissues brought on by underlying illnesses may affect the material's physical and chemical properties. Acidic environments demonstrate a reduction in MTA's hardness[97,98] and sealing ability.[99] However, no studies have explored the strength of dentin's bond with MTA in an acidic environment. The purpose of this observation was to find out how strong the push-out binding is between intraarticular dentin and MTA when they are exposed to different acidic pH levels.

10.4 FUTURE DIRECTIONS IN RAMAN SPECTROSCOPY FOR DENTAL MATERIALS

10.4.1 Portable Raman Devices for Clinical Use

Raman spectroscopy's rapid and non-destructive study of both inorganic and organic substances is one of its key advantages for criminal and geotechnical field investigations. For instance, law enforcement, security, and first responders must act swiftly and precisely when detecting explosives, street drugs, and other substances of forensic interest. For these reasons, the portability of the instrumentation is essential.[100–103] Certain applications require portable technology due to the importance of measuring the sample at remote locations. Micro-sampling (larger than or equivalent to a square millimeter) of tissue or biological fluids can be used to perform clinical RS measurements, where proteins, lipids, and nucleic acids account for the majority of the signal within the stimulated tissue volume. Spectral analysis can infer changes in pathological tissue diseases, which often precede and correlate with metabolic changes. Customization is necessary for every use of the Raman system, from small systems for space missions[104–106] to sampling probes that are robust to pressure and temperature for industrial monitoring and geological research.[107,108] Biological tissue inspection and diagnostics are increasingly using Raman spectroscopy.[109–111] For a system of this kind, simplicity, portability, and a satisfactory signal-to-noise ratio may be the most crucial design factors. Even individuals with limited or no knowledge of spectroscopy should be able to operate the medical equipment due to its small size and easy-to-use controls. Furthermore, many malignancies initiate in specific tissue locations. For instance, the stratum basale, the single layer of cells that separates the epidermis from the dermis, is known to be the source of many skin malignancies. To improve the effectiveness of Raman spectroscopy for early diagnosis, it might be necessary to measure the Raman signal from this layer independently of the surrounding tissue. Physical cutting is no longer necessary because of confocal Raman microspectroscopy, which records Raman signals from precise depths and positions and measures small, fixed areas. "Optical sectioning" has recently been utilised to investigate the composition of artery walls[112,113] the dynamics of skin moisture,[114,115] and the biochemistry of eye tissue.[116] Unfortunately, most of the papers that talk about how to build and use Confocal Raman Microspectrometers (CRMs) describe big, stationary machines that are difficult to move around in a clinical setting.[117–119] Additionally, the majority of commercial CRM systems rely on microscope frames with epi-illumination, which necessitates setting the sample up on an upright or inverted stage. Nowadays, only a few independent businesses sell portable Raman spectrometers,[120–123] as shown in Figure 10.6.[124] Their unique advantages over conventional analytical techniques, such as their capacity to extract the fingerprint spectrum of materials without damaging them or without extra sample preparation, are making them increasingly popular.[125] Therefore, there is no need for labor-intensive, equipment-heavy analysis procedures when using these devices in field chemical assessments, environmental protection, forensic science, and counter-terrorism.[126] Raman spectroscopy is used in a variety of applications, including the field identification of forensic traces at crime scenes, fast identification of unknown substances at airports or government buildings (like post or customs offices), field identification of smuggled or controlled substances (like alcohol, fuels, or drugs), and pollution detection in water or soil. The technical needs of portable systems and looking at the spectral

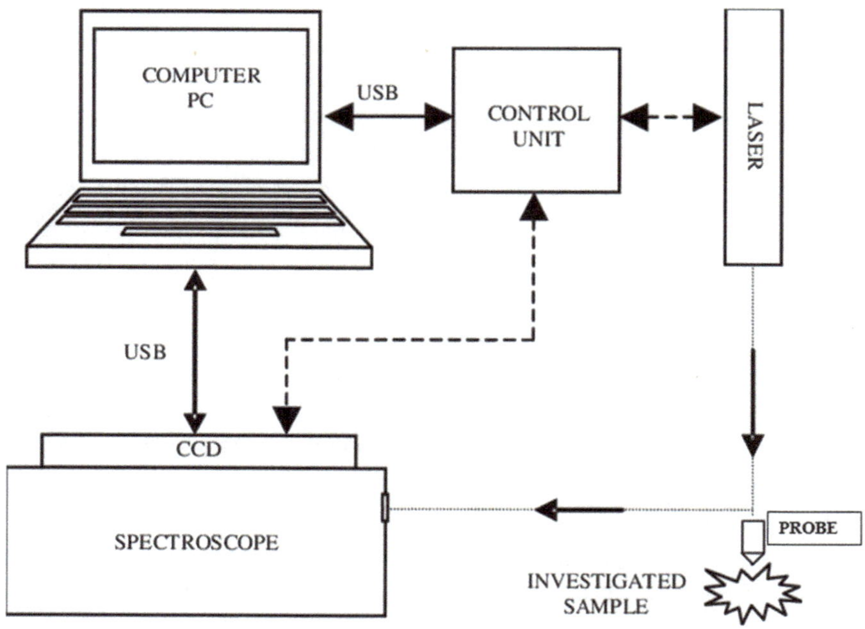

FIGURE 10.6 Block diagram of a portable Raman spectrometer.[124]

features of measurements taken in living things show that Raman spectroscopy is moving closer to being used in medicine.

10.4.2 Integration with CAD/CAM Technologies in Dentistry

In dentistry, the terms "CAD/CAM" refer to the method of creating an indirect restoration using computer-aided design (CAD) and computer-assisted machine milling (CAM).[127] Duret and associates created this initial system in 1971, but several issues, including inadequate computer resources, accuracy in digitizing, and other issues, hindered its extensive use.[128] Since the advent of information about CAD/CAM, the area of odontology has experienced substantial novelty. This instrument has been used in dentistry for both therapeutic and diagnostic procedures. Knowledge of CAD/CAM makes a fully digital blueprint possible. Furthermore, certain protocol reports have indicated a possible clinical consistency, and patients have reported very excellent outcomes with this equipment.[129,130] Retrospective research also suggested that CAM functionalities reduced management time.[131] Computer-aided impression taking, digital tooth shade determination, chairside and laboratory restoration fabrication, and a variety of digital imaging formats such as dental radiographs, tomography, and cone-beam computed tomography, among others, utilize digital Dental education is also included in digital learning opportunities. We must update the dental curriculum digitally, incorporating virtual technologies and software interpretations for dental operations.[132] Undergraduate dentistry students today differ from their predecessors in that they belong to a digitally intermediated generation and are familiar with virtual experience and simulation. Typically, colleges expect undergraduate dentistry students to authorize and even select digital techniques due to their familiarity with computers in a virtual environment.[133–134] Furthermore, almost all colleges are looking into more capable ways to teach students, including digital abilities.[135] Dental schools have also reported on the results of having their undergraduate students operate a variety of instruments. Consequently, research has demonstrated a correlation between the efficiency of dentistry student implementation and factors such as time, efficacy, and faculty understanding.[136,137] New high-strength ceramic materials for dental devices have recently been introduced in response to the increased need for dental materials that are both safe and visually pleasing.[138] It has been predicted that these materials will negatively impact conventional dental processing technology; as a result, new, advanced dental processing creating

CAD/CAM technologies is one approach to overcome. When we began research and development in the 1980s, many believed that the design and processing of dental equipment using CAD/CAM technology would be simpler and easier than for industrial items. Recent developments in the expansion of CAD/CAM systems for use in industrial rapid prototyping have made it possible to significantly improve software and reduce the size and cost of gear. These developments have led to the adoption of CAD/CAM by many other industries, including the dental and medical fields. Dental practitioners have discovered that CAD/CAM technology is very helpful in that it eliminates the need for conventional analog dental laboratory operations and makes it possible to fabricate personalized, patient-specific prostheses and restorations.

10.5 CONCLUSION

Raman spectroscopy is quickly changing from an experimental way to study biological molecules to a therapeutic screening tool for instantaneous evaluation of oral tissue in the fields of medicine and orthodontics. This chapter's goals are to list the different uses for Raman spectroscopy, assess the state of the field now, and consider potential future developments in dentistry. It provides extensive information on polymorphism, crystal orientations, and structural organization. Furthermore, it evaluates the interactions between molecules and the impact of stress and environmental factors on samples. Because it is very specific to biochemicals, doesn't react negatively with water, is simple to assess spectral parameters for, can operate in the near-infrared, and can deliver potentials to living things using fiber optics, this technology is becoming more and more significant in biomedical research. Raman spectroscopy is a useful and versatile technique for analyzing the microstructure and nanostructure properties of dental materials. The ability to discern between different phases, detect minute chemical changes, and analyze component distribution at the micro- and nanoscale levels has greatly advanced our understanding of dental materials. It has made it possible to thoroughly examine biomaterials like ceramics, dental composites, and others, resulting in the creation of materials with improved durability, mechanical properties, and biocompatibility. Despite these incredible developments, standardizing Raman spectroscopic techniques for use in dentistry remains challenging. When combined with other cutting-edge imaging methods like SEM and X-rays, Raman spectroscopy may enable a future study to go deeper and explore in more dimensions. Additionally, developments in portable

Raman technology could pave the way for in-situ monitoring of dental procedures and real-time diagnostics in clinical settings. Ultimately, we expect to see more studies on dental materials using Raman spectroscopy.

REFERENCES

1. Movasaghi, Z., Rehman, S., Rehman, I.U., Fourier Transform Infrared (FTIR) spectroscopy of biological tissues, *Applied Spectroscopy Reviews* 43, no. 2 (2008):134–179.
2. Miyazaki, M., Onose, H., Moore, B.K. Analysis of the dentin-resin interface by use of laser Raman spectroscopy, *Dental Materal* 18, no. 8 (2002):576–580.
3. Ko, A.C.T., Choo-Smith, L.P., Hewko, M., and Sowa, M.G. Detection of early dental caries using polarized Raman spectroscopy, *Optics Express* 14, no. 1 (2006):203–215.
4. Venkatakrishna, K., Kurien, J., Pai, M.K. et al. Optical pathology of oral tissue: A Raman spectroscopy diagnostic method. *Current Science* 80, no 5 (2001):665–669.
5. Kah, J.C.Y., Kho, K.W., Lee, C.G.L. et al., Early diagnosis of oral cancer based on the surface plasmon resonance of gold nanoparticles, *International Journal of Nanomedicine* 2, no. 4 (2007):785–798.
6. Krafft, C., Sergo, V., Biomedical applications of Raman and infrared spectroscopy to diagnose tissues, *Spectroscopy* 20 (2006):195–218. 10.1155/2006/738186
7. Boskey, A.L., Mendelsohn, R. Infrared spectroscopic characterization of mineralized tissues, *Vibrational Spectroscopy* 38, no. 1–2 (2005):107–114. 10.1016/j.vibspec.2005.02.015
8. Niklas, C., Wackerbarth, H., Ctistis, G. A short review of cavity-enhanced raman spectroscopy for gas analysis, *Sensors* 21, no. 5 (2021):1698. doi:10.3390/s21051698
9. Ko, A.C.T., Hewko, M., Sowa, M. G. Early dental caries detection using a fiber-optic coupled polarization-resolved Raman spectroscopic system, *Optics Express* 16, no. 9 (2008):6274–6284. 10.1364/oe.16.006274
10. Vargas-Koudriavtsev, T., Durán-Sedó, R. et al., Effect of tooth-bleaching agents on phosphate concentration in dental enamel by means of Raman spectroscopy, *Revista Odontológica Mexicana* 19, no. 4 (2015):228–235. doi:10.1016/j.rodmex.2015.10.013
11. Tsuda, H., Arends, J., Raman spectroscopy in dental research: A short review of recent studies, *Advances in Dental Research*, 11, no. 4 (1997):539–547. 10.1177/08959374970110042301
12. Monteiro, M., Chasqueira, F., Sofia, P., Raman spectroscopy in the characterization of carious dental tissues. *Spectroscopy Europe/World* 30, no. 3, (2018):11–14.
13. Toporski, J., Dieing. T., *Raman Microscopy (Confocal). I Encyclopedia of Geobiology*; Springer: Dordrecht, The Netherlands, (2011):754–761.
14. Zhang, Y., Ren, L., Qi, W. et al., Raman spectroscopy: A potential diagnostic tool for oral diseases, *Frontiers in Cellular Infection Microbiology* 12, (2022):775236. 10.3389/fcimb.2022.775236

15. Eberhardt, E., Stiebing, C., Matthaus, C., Schmitt, M., Popp, J., Advantages and limitations of Raman spectroscopy for molecular diagnostics: An update, *Expert Review of Molecluar Diagnostic* 15, no. 6 (2015):773–787. 10.1586/14737159.2015.1036744
16. Ferraro, J.R., Nakamoto, K., *Introductory Raman Spectroscopy*, 2nd ed.; Elsevier: Amsterdam, The Netherlands, 2003; ISBN 978-0-12-254105-6.
17. Vo-Dinh, T., *Biomedical Photonics Handbook*; CRC Press: Boca Raton, FL, USA, 2003. doi:10.1201/9780203008997
18. Matousek, P., Morris, M., *Emerging Raman Applications and Techniques in Biomedical and Pharmaceutical Fields*; Springer: Berlin/Heidelberg, Germany, 2010; ISBN 978-3-642-02648-5. 10.1007/978-3-642-02649-2
19. Sun, Q., Zhang, L., Bai, R., et al., Recent progress in antimicrobial strategies for resin-based restoratives, *Polymers* 13, no. 10 (2021):1590. doi:10.3390/polym13101590
20. Albelasy, E. H., Hamama, H. H., Chew, H.P., Montaser, M., Mahmoud, S. H., Secondary caries and marginal adaptation of ion-releasing versus resin composite restorations: a systematic review and meta-analysis of randomized clinical trials, *Scientific Reports* 12, no. 1 (2022):19244. doi:10.1038/s41598-022-19622-6
21. Cavalcante, L.M., Ramos, A. B., Silva, D.C., et al., Thiourethane-functionalized fillers: biological properties and degradation resistance, *Brazilian Oral Research* 35, (2020):e018. doi:10.1590/1807-3107bor-2021.vol35.0018
22. Sezen, M., Focused Ion Beams (FIB): Novel Methodologies and Recent Applications for Multidisciplinary Sciences. In: Janecek, M. (Ed.), *Modern Electron Microscopy In Physical And Life Sciences*. Intech, Rijeka, (2016): 121–140.
23. Duckworth, R.M., 2016. The Teeth and Their Environment, Physical, Chemical and Biochemical Influences. *Monographs in Oral Science*. Karger Ag, Basel.
24. Kinney, J. H., Nalla, R. K., Pople, J. A., Breunig, T. M., Ritchie, R. O., Age-related transparent root dentin: mineral concentration, crystallite size, and mechanical properties. *Biomaterials* 26, no. 16 (2005):3363–3376. 10.1016/j.biomaterials.2004.09.004
25. Porter, A. E., Nalla, R. K., Minor, A., et al., A transmission electron microscopy study of mineralization in age-induced transparent dentin. *Biomaterials* 26, no. 36 (2005):7650–7660. 10.1016/j.biomaterials.2005.05.059
26. Jones, F. H., Teeth and bones: Applications of surface science to dental materials and related biomaterials, *Surface Science Reports* 42, no. 3 (2001):75–205.
27. Nalla, R. K., Porter, A. E., Daraio, C., Minor, A. M., Radmilovic, V., Stach, E. A., et al., Ultrastructural examination of dentin using focused ion-beam cross-sectioning and transmission electron microscopy, *Micron* 36, no. 7–8 (2005):672–680. 10.1016/j.micron.2005.05.011
28. Marshall, S. J., Balooch, M., Habelitz, S., Balooch, G., Gallagher, R., Marshall, G.W., The dentin–enamel junction – a natural, multilevel interface, *Journal of the European Ceramic Society* 23, no. 15 (2003):2897–2904. 10.1016/S0955-2219(03)00301-7

29. Kinney, J. H., Pople, J. A., Marshall, G. W., Marshall, S. J., Collagen orientation and crystallite size in human dentin: a small angle X-ray scattering study, *Calcified Tissue International* 69, no. 1 (2001):31–37. 10.1007/s00223-001-0006-5
30. Gawda, H., Sekowski, L., Treracz, H. In vitro examination of human teeth using ultrasound and X-ray diffraction. *Acta of Bioengineering and Biomechanics* 6, no. 1 (2004):41–49.
31. Kinney, J. H., Marshall, S. J., Marshall, G. W., The mechanical properties of human dentin: a critical review and re-evaluation of the dental literature. *Critical Review of Oral Biology and Medicine* 14, no. 1 (2003):13–29. 10.1177/154411130301400103
32. Marshall, G. W., Dentin: microstructure and characterization, *Quintessence International* 24, no. 9 (1993):606–617.
33. Xu, C., Yao, X., Walker, M. P., Wang, Y., Chemical/molecular structure of the dentin–enamel junction is dependent on the intra tooth location, *Calcified Tissue International* 84, no. 3 (2009):221–228. 10.1007/s00223-008-9212-8
34. Gher, M. J., Dunlap, R., Anderson, M., Kuhl, L., Clinical survey of fractured teeth, *Journal of American Dental Association* 114, (1987):174–177.
35. Gilman, J. J., Chemistry and Physics of Ceramics. *Scientific American*, 217, no. 4 (1967): 120–130. https://doi.org/10.1038/scientificamerican1067-120
36. Guazzato, M., Albakry, M., Ringer, S. P., Swain, M. V., Strength, fracture toughness and microstructure of a selection of all-ceramic materials. Part II. Zirconia-based dental ceramics, *Dental Materials* 20, no. 5 (2004):449–456. 10.1016/j.dental.2003.05.002
37. Kelly, J. R., Nishimura, I., Campbell, S. D., Ceramics in dentistry: Historical roots and current perspectives, *The Journal of Prosthetic Dentistry* 75, no. 1 (1996):18–32. 10.1016/s0022-3913(96)90413-8
38. Seha, M. M., Pamuk, S., Balkaya, M. C., Akgungor, G., Effect of Tuf-Coat on Feldspathic porcelain materials. *Journal of Oral Rehabilitation* 32, no. 1 (2005):39–45. 10.1111/j.1365-2842.2004.01367.x
39. Van Dijken, J. W., Ormin, A., Olofsson, A. L., Clinical performance of pressed ceramic insulted with resin-modified glass ionomer and auto polymerizing resin composite cement, *The Journal of Prosthetic Dentistry* 82, no. 5 (1999):529–535. 10.1016/s0022-3913(99)70051-x
40. Scherrer, S. S., Kelly, J. R., Quinn, G. D., Xu, K., Fracture toughness (KIc) of dental porcelain determined by fractographic analysis, *Dental Materials* 15, no. 5 (1999):342–348. 10.1016/s0109-5641(99)00055-x
41. Barnfather, K. D., Brunton, P. A., Restoration of the upper dental arch using Lava all ceramic crown and bridgework, *British Dental Journal* 202, no. 12 (2007):731–735. 10.1038/BDJ.2007.534
42. Lacefield, W.R. Current status of ceramic coatings for dental implants. *Implant Dentistry* 7, no. 4 (1998):315–322. 10.1097/00008505-199807040-00010
43. Touati, B., Miara, P., and Nathanson, D.. Esthetic dentistry and ceramic restorations. *British Dental Journal* 186 (1999): 533.

44. Callister, W., *Materials Science and Engineering: An introduction.* 3rd ed. New York: John Wiley and Sons 1994.
45. DeMaar, F. E., Historically, when and by whom was silver amalgam introduced? *Scientific and Education Bulletin* 5, no. 1 (1972):23–26.
46. Black, GV. The physical properties of the silver-tin amalgams, *Dental Cosmos* 38, no. (1896):965–992. 10.14219/jada.archive.1978.0100
47. Clarkson, T. W., Magos, L., Myers, G.J., The toxicology of mercury— current exposures and clinical manifestations, *The New England Journal of Medicine* 349, no. 18 (2003):1731–1737. 10.1056/NEJMra022471
48. Jones, D.W., The enigma of amalgam in dentistry. *Journal Canadian Dental Association* 59, no. 2 (1983):155–166.
49. Langan, D. C., Fan, P. L., Hoos, A. A., The use of mercury in dentistry: A critical review of the recent literature. *Journal of the American Dental Association* 115, no. 6 (1987):867–880. 10.14219/jada.archive.1987.0270
50. Mai, E. K., Qasem, D. A., Ridwaan, O., Factors relating to usage patterns of amalgam and resin composite for posterior restorations—A prospective analysis, *Journal of Dentistry* 42, no. 7 (2014):785–792. 10.1016/j.jdent.2014.04.010
51. Rathore, M., Singh, A., Pant, V.A. The dental amalgam toxicity fear: A myth or actuality, *Toxicology International* 19, no. 2 (2012):81–88. 10.4103/0971-6580.97191
52. United Nations Environment Programme—Chemicals. *Global Mercury Assessment Inter-Organization Programme for the Sound Management of Chemicals; Inter-Organization Programme for the Sound Management of Chemicals—Inter-Organization Programme for the Sound Management of Chemicals—IOMC*: Geneva, Switzerland, 2002.
53. Cenci, M. S., Piva, E., Potrich, F., Formolo, E., Demarco, F. F., Powers, J. M. Microleakage in bonded amalgam restorations using different adhesive materials, *Brazilian Dental Journal* 15, no. 1(2004):13–18. 10.1590/s0103-64402004000100003
54. Beyth, N., Domb, A. J., Weiss, E.I., An in vitro quantitative antibacterial analysis of amalgam and composite resins. *Journal of Dentistry* 35, no. 3 (2007):201–206. 10.1016/j.jdent.2006.07.009
55. Goldman, A., Frencken, J. E., De Amorim, R. G., Leal, S. C., Replacing amalgam with a high-viscosity glass-ionomer in restoring primary teeth: A cost-effectiveness study in Brasilia, *Journal of Dentistry* 70, (2018):80–86. 10.1016/j.jdent.2017.12.012
56. ISO-Standards ISO 1559. *Dental Materials—Alloys for Dental Amalgam; International Organization for Standardization*: Geneva, Switzerland, 1995.
57. Tchounwou, P. B., Ayensu, W. K., Ninashvili, N., Sutton, D., Environmental exposure to mercury and its toxicopathologic implications for public health, *Environmental Toxicology* 18, no. 3 (2003):149–175. 10.1002/tox.10116
58. Dunne, S. M., Grainsford, I. D., Wilson, N. H., Current materials and techniques for direct restorations in posterior teeth. *Part I: Silver amalgam, International Dental Journal* 47, no. 3 (1997):123–136. 10.1002/j.1875-595x.1997.tb00777.x

59. Hartung, R., Dinman G. D., *Environmental mercury contamination/Rolf Hartung*, Bertram D. Dinman, editors. Michigan: Ann Arbor Science Publishers, 1972.
60. Bayne, S. C., Heymann, H. O., Swift Jr, E. J., Update on dental composite restorations. *Journal of the American Dental Association* 125, no. 6 (1994):687–701.10.14219/jada.archive.1994.0113
61. Jandt, K. D., Sigusch, B.W., Future perspectives of resin-based dental materials, *Dental Materials: Official Publication of the Academy of Dental Materials* 25, no. 8 (2009):1001–1006. 10.1016/j.dental.2009.02.009
62. Ferracane, J. L., Resin composite–state of the art, *Dental Material* 27, no. 1 (2011):29–38.
63. Janus, J., Fauxpoint, G., Arntz, Y., Pelletier, H., Etienne, O., Surface roughness and morphology of three nanocomposites after two different polishing treatments by multi-technique approach. *Dental Materials: Official Publication of the Academy of Dental Materials* 26, no. 5 (2010):416–425. 10.1016/j.dental.2009.09.014
64. Musanje, L., Darvell, B., Aspects of water sorption from the air, water and artificial saliva in resin composite restorative materials, *Dental Materials: Official Publication of the Academy of Dental Materials* 19, no. 5 (2003):414–422. 10.1016/s0109-5641(02)00085-4
65. Teughels, W., Van Assche, N., Sliepen, I., Quirynen, M. Effect of material characteristics and / or surface topography on biofilm development. *Clin. Oral Implantol. Res* 17, no. 2 (2006):68–81. 10.1111/j.1600-0501.2006.01353.x
66. Watanabe, T., Miyazaki, M., Takamizawa, T., Influence of polishing duration on surface roughness of resin composites, *Journal of Oral Science* 47, no. 1 (2005):21–25. 10.2334/josnusd.47.21
67. Sharma, S., Cross, S. E., Hsueh, C., Wali, R. P., Stieg, A. Z., Gimzewski, J. K., et al., Nanocharacterization in dentistry, *International Journal of Molecular Sciences* 11, no. 6 (2010):2523–2545. doi:10.3390/ijms11062523
68. Uskokovi, C. V., Bertassoni, L. E., Nanotechnology in dental sciences: Moving towards a finer way of doing dentistry *Materials (Basel)* 3, no. 3 (2010):1674–1691. 10.3390/ma3031674
69. Salerno, M., Patra, N., Cingolani, R., Use of ionic liquid in fabrication, characterization, and processing of anodic porous alumina, *Nanoscale Research Letters* 4, no. 8 (2009):865–872. 10.1007/s11671-009-9337-3
70. Rezvani, M. B., Atai, M., Hamze, F., Hajrezai, R., The effect of silica nanoparticles on the mechanical properties of fiber-reinforced composite resins, *Journal of Dental Research, Dental Clinics, Dental Prospects* 10, no. 2 (2016):112–117. 10.15171/joddd.2016.018
71. Ramesh, S., Teja, K., Priya, V., Regulation of matrix metalloproteinase-3 gene expression in inflammation: A molecular study, *Journal of Conservative Dentistry* 21, no. 6 (2018):592–596. 10.4103/jcd.jcd_154_18
72. Jain, R., Tiwari, D.C., Preeti, P., Highly selective and sensitive graphene-based electrochemical sensor for quantification of receptor agonist rizatriptan, *Electroanalysis* 25 no. 6 (2013):1363–1367. doi:10.1002/elan.201300003

73. Jain, R., Tiwari, D. C., Pnadey Preeti., Electrocatalytic quantifaction of anti-tubercular drug ethionamide in sodium lauryl sulfate, *Journal of Molecular Liquids* 198 no. (2014):364–368. doi:10.1016/j.molliq.2014.07.025
74. Buonocore, M. G., A simple method of increasing the adhesion of acrylic filling materials to enamel surfaces, *Journal of Dental Research* 34, no. 6 (1955):849–853. 10.1177/00220345550340060801
75. Ferracane, J. L., Mitchem, J. C., Condon, J. R., Todd, R., Wear and marginal breakdown of composites with various degrees of cure, *Journal of Dental Research* 76, no. 8 (1997):1508–1516. 10.1177/00220345970760081401
76. Ferracane, J. L., Berge, H. X., Fracture toughness of experimental dental composites aged in ethanol, *Journal of Dental Research* 74, no. 7 (1995):1418–1423. 10.1177/00220345950740071501
77. Rueggeberg, F. A., Craig, R. G., Correlation of parameters used to estimate monomer conversion in a light-cured composite, *Journal of Dental Research* 67, no. 6 (1988):932–937. 10.1177/00220345880670060801
78. Ferracane, J., Developing a more complete understanding of stresses produced in dental composites during polymerization, *Dental Materials: Official Publication of the Academy of Dental Materials* 21, no. 1 (2005):36–42. 10.1016/j.dental.2004.10.004
79. Ferracane, J. L., Greener, E. H., The effect of resin formulation on the degree of conversion and mechanical properties of dental restorative resins, *Journal of Biomedical Materials Research* 20, no. 1 (1986):121–131. 10.1002/jbm.820200111
80. Johnston, W. M., Reisbick, M. H., Color and translucency changes during and after the curing of esthetic restorative materials, *Dental Materials* 13, no. 2 (1997):89–97. 10.1016/s0109-5641(97)80017-6
81. Leinfelder, K. F., Radz, G. M., Nash, R. W., A report on a new condensable composite resin, *Compendium of Continuing Education in Dentistry* 19, no. 3 (1998):230–237.
82. Jandt, K. D., Sigusch, B. W., Future perspectives of resin-based dental materials, *Dental Materials: Official Publication of the Academy of Dental Materials* 25, no. 8 (2009):1001–1006. 10.1016/j.dental.2009.02.009
83. Chen, M. H., Update on dental nanocomposite. *Journal of Dental Research* 89, no. 6 (2010):540–560. 10.1177/0022034510363765
84. Burke, F. J., Amalgam to tooth-colored materials-implications for clinical practice and dental education: governmental restrictions and amalgam-usage survey results. Journal of Dentistry 32, no. 5 (2004):343–350. 10.1016/j.jdent.2004.02.003
85. Darwell, B.W. (1971) *MSc thesis*, University of Wales.
86. Nomoto, S., Ano, M., & Onose, H. Micro-probe for measurement of corrosion potential of metallic restorations in mouth. *Journal of Dental Research* 58, no. 7, (1979):1688–1690. doi:10.1177/00220345790580070601
87. Westerhoff, B., Darwish, M., & Holze, R. A comparative study of the electrochemical corrosion behaviour of dental amalgams. *Journal of Applied Electrochemistry* 22, no. 12 (1992):1142–1146. doi:10.1007/bf01297415

88. Nakagawa, M., Matsuya, S., Ohta, M. Effect of microstructure on the corrosion behaviour of dental gold alloys. *Journal of Materials Science: Materials in Medicine* 3, no. 2 (1992):114–118. doi:10.1007/bf00705278
89. Uhlig, H. H. (1967) *Corrosion and Corrosion Control*, 4th edn, pp. 6–112. John Wiley & Sons, New York.
90. Rabald, E. (1968) *Corrosion Guide*, 2nd edn, pp. 199–301. Elsevier Publishers, Amsterdam.
91. Geis-Gerstorfer, J., In vitro corrosion measurements of dental alloys. *Journal of Dentistry* 22, no. 4, (1994):247–251. doi:10.1016/0300-5712(94)90124-4
92. Kaesche, H. (1979) *Die Korrosion der Metalle*, 2nd edn, pp. 10–115. Springer Verlag, Berlin.
93. Aksut, A. A., Bayramolu, G., The effect of propargyl alcohol on the corrosion of pure aluminium and aluminium alloys in aqueous solutions, *Corrosion Science* 36, no. 3 (1994):415–422. doi:10.1016/0010-938x(94)90034-5
94. Aksüt, A. A., Bilgiç, S., The effect of amino acids on the corrosion of nickel in H2SO4. *Corrosion Science* 33, no. 3 (1992):379–387. doi:10.1016/0010-938x (92)90067-d
95. Nekoofar, M. H., Namazikhah, M. S., Sheykhrezae, M. S., Mohammadi, M. M., Kazemi, A., Aseeley, Z., Dummer, P. M. H., pH of pus collected from periapical abscesses. *International Endodontic Journal* 42 no. 6 (2009):534–538. doi:10.1111/j.1365-2591.2009.01550.x
96. Torabinejad, M., & Chivian, N. Clinical applications of mineral trioxide aggregate. *Journal of Endodontics* 25, no. 3, (1999):197–205. doi:10.1016/s0099-2399(99)80142-3
97. Lee, Y. L., Lee, B. S., Lin, F. H., et al., Effects of physiological environments on the hydration behavior of mineral trioxide aggregate, *Biomaterials* 25, no. 5 (2004):787–793. doi:10.1016/s0142-9612(03)00591-x
98. Namazikhah, M. S., Nekoofar, M. H., Sheykhrezae, M. S., et al. The effect of pH on surface hardness and microstructure of mineral trioxide aggregate, *International Endodontic Journal* 41, no. 2 (2007):108–116. doi:10.1111/j.1365-2591.2007.01325.x
99. Saghiri, M. A., Lotfi, M., Saghiri, A. M., Fatemi, A., Ranjkesh, B., et al., Effect of pH on sealing ability of white mineral trioxide aggregate as a root-end filling material. *Journal of Endodontics 34*, no. 10 (2008):1226–1229. doi:10.1016/j.joen.2008.07.01
100. Edwards, H. G. M., Munshi, T., Diagnostic Raman spectroscopy for the forensic detection of biomaterials and the preservation of cultural heritage. *Analytical and Bioanalytical Chem*istry 382, (2005):1398–1406. doi:10.1007/s00216-005-3271-8
101. Hargreaves, M. D., Page, K., Munshi, T., Tomsett, R., Lynch, G., Edwards, H. G. M. Analysis of seized drugs using portable Raman spectroscopy in an airport environment-a proof of principle study. *Journal of Raman Spectroscopy* 39, no. 7 (2008):873–880. doi:10.1002/jrs.1926
102. Zieba-Palus, J., Borusiewicz, R., Kunicki, M., PRAXIS—combined μ-Raman and μ-XRF spectrometers in examining forensic samples. *Forensic Science International* 175, no. 1 (2008):1–10. doi:10.1016/j.forsciint.2007.04.230

103. Moore, D. S., Scharff, R.J., Portable Raman explosives detection. *Analytical and Bioanalytical Chemistry* 393, no. 6-7 (2009):1571–1578. doi:10.1007/s00216-008-2499-5
104. Dickensheets, D. L., Wynn-Williams, D. D., Edwards, H. G. M., Schoen, C., Crowder, C., & Newton, E. M. A novel miniature confocal microscope/Raman spectrometer system for biomolecular analysis on future Mars missions after Antarctic trials. *Journal of Raman Spectroscopy*31 no. 7, (2000):633–635. doi:10.1002/1097-4555(200007)31:7<633::aid-jrs620>3.0.co;2-r
105. Edwards, H. G., Newton, E. M., Dickensheets, D. L., & Wynn-Williams, D. D. Raman spectroscopic detection of biomolecular markers from Antarctic materials: evaluation for putative Martian habitats, *Spectrochimica Acta Part A: Molecular and Biomolecular Spectroscopy* 59, no. 10, (2003):2277–2290. doi:10.1016/s1386-1425(03)00071-4
106. Wang, A., Haskin, L. A., Cortez, E., Prototype Raman spectroscopic sensor for in situ mineral characterization on planetary surfaces, *Applied Spectros*copy 52, no. 4 (1998):477–487. doi:10.1366/0003702981943842
107. Gorbaty, Y. E., Bondarenko, G. V. High-pressure high-temperature Raman cell for corrosive liquids. *Review of Scientific Instruments* 66, no. 8 (1995):4347–4349. doi:10.1063/1.1145326
108. Schiferl, D., Sharma, S. K., Cooney, T. F., Wang, S. Y., Mohanan, K., Multichannel Raman spectrometry system for weakly scattering materials at simultaneous high pressures and high temperatures. *Review of Scientific Instruments* 64, no. 10 (1993):2821–2827. doi:10.1063/1.1144368
109. Hanlon, E. B., Manoharan, R., Koo, T.-W., Shafer, K. E., Motz, J. T., Fitzmaurice, F., M. S. et al., Prospects for vivo Raman spectroscopy, *Physics in Medicine and Biology* 45, no. 2 (2000): R1–R59. doi:10.1088/0031-9155/45/2/201
110. Mahadevan-Jansen, Anita and Richards-Kortum, Rebecca R. "Raman spectroscopy for the detection of cancers and precancers," *Journal of Biomedical Optics* 1, no. 1 (1996). doi:10.1117/12.227815
111. Stone, N., Kendall, C., Shepherd, N., Crow, P., & Barr, H. Near-infrared Raman spectroscopy for the classification of epithelial pre-cancers and cancers, *Journal of Raman Spectroscopy* 33, no. 7 (2002):564–573. doi:10.1002/jrs.882
112. Buschman, H. P., Deinum, G., Motz, J. T., Fitzmaurice, M., Kramer, J. R., van der Laarse, A., et al., Raman microspectroscopy of human coronary atherosclerosis: Biochemical assessment of cellular and extracellular morphologic structures in situ. *Cardiovascular Pathology* 10, no. 2 (2001):69–82. doi:10.1016/s1054-8807(01)00064-3
113. Buschman, H. P., Marple, E. T., Wach, M. L., Bennett, B., Bakker Schut, T. C., Bruining, H. A., et al., In vivo determination of the molecular composition of artery wall by intravascular raman spectroscopy, *Analytical Chemistry* 72, no. 16 (2000):3771–3775. doi:10.1021/ac000298b
114. Caspers, P. J., Lucassen, G. W., Carter, E. A., Bruining, H. A., Puppels, G. J., In vivo confocal Raman microspectroscopy of the skin: Noninvasive determination of molecular concentration profiles, *Journal of Investigate Dermatology* 116, no. 3 (2001):434–442. doi:10.1046/j.1523-1747.2001.01258.x

115. Chrit, L., Bastien, P., Sockalingum, G. D., Batisse, D., Leroy, F., Manfait, M., & Hadjur, C. An in vivo randomized study of human skin moisturization by a new confocal Raman fiber-optic microprobe: Assessment of a glycerol-based hydration cream, *Skin Pharmacology and Physiology* 19, no. 4 (2006):207–215. doi:10.1159/000093116
116. Erckens, R. J., Motamedi, M., March, W. F., Wicksted, J. P. Raman spectroscopy for non-invasive characterization of ocular tissue: Potential for detection of biological molecules. *Journal of Raman Spectroscopy* 28, no. 5 (1997):293–299. doi:10.1002/(sici)1097-4555(199705)28:5<293::aid-jrs47>3.0.co;2-0
117. Caspers, P. J., Lucassen, G. W., Bruining, H. A., Puppels, G. J. Automated depth-scanning confocal Raman microspectrometer for rapidin vivo determination of water concentration profiles in human skin. *Journal of Raman Spectroscopy* 31, no. 8–9 (2000):813–818. doi:10.1002/1097-4555(200008/09)31:8/9<813::aid-jrs573>3.0.co;2-7
118. Chrit, L., Hadjur, C., Morel, S., Sockalingum, G., Lebourdon, G., Leroy, F., (2005). In vivo chemical investigation of human skin using a confocal Raman fiber optic microprobe. *Journal of Biomedical Optics* 10, no. 4 (2005):044007. doi:10.1117/1.2003747
119. Bakker Schut, T. C., Caspers, P. J., Puppels, G. J., Nijssen, A., Heule, F., Neumann, M. H., et al. Discriminating basal cell carcinoma from its surrounding tissue by Raman spectroscopy. *Journal of Investigative Dermatology* 119, no. 1 (2002):64–69. doi:10.1046/j.1523-1747.2002.01807.x
120. Dickensheets, D. L., Wynn-Williams, D. D., Edwards, H. G. M., Schoen, C. Crowder, C., Newton E. M., A novel miniature confocal microscope/Raman spectrometer system for biomolecular analysis on future Mars missions after Antarctic trials, *Journal of Raman Spectroscopy* 31, no. 7 (2000):633–635. doi:10.1002/1097-4555(200007)31:7<633::AID-JRS620>3.0.CO;2-R
121. Vítek, P., Ali, E. M. A., Edwards, H. G. M., Jehlička, J., Cox, R., & Page, K., Evaluation of portable Raman spectrometer with 1064nm excitation for geological and forensic applications, *Spectrochimica Acta Part A: Molecular and Biomolecular Spectroscopy* 86 (2012):320–327. doi:10.1016/j.saa.2011.10.043
122. Lieber, C., & Mahadevan-Jansen, A., Development of a handheld Raman micro spectrometer for clinical dermatologic applications, *Optics Express* 15, no. 19 (2007):11874–11882. doi:10.1364/oe.15.011874
123. Gnyba, M., Smulko, J., Kwiatkoswski, A., Wierzba, P., Portable Raman spectrometer design rules and applications, *Bulletin of the Polish Academy of Sciences Technical Sciences* 59, no. 3 (2011):325–329. 10.2478/v10175-011-0040-z
124. Kwiatkowski, A., Gnyba, M., Janusz M. Smulko J., Wierzba, P., Algorithms of chemicals detection using Raman spectra, *Metrology and Measurement Systems* 17 (2010):549–559.
125. Lam, T., A new era in affordable Raman spectroscopy, *Raman Technology for Today's Spectroscopists* 3 (2004):30–37.
126. McCreery, R. L., *Raman Spectroscopy for Chemical Analysis*, John Wiley & Sons, New York, 2000.

127. Correia A. R. M., Sampaio Fernandes J. C. A., Cardoso J. A. P., Leal-Da-Silva C. L. C. CAD-CAM: Informatics applied to fixed prosthodontics, *Revista de Odontologia da UNESP* 35, no. 2 (2006):183–189.
128. Miyazaki, T., Hotta, Y., CAD/CAM systems available for the fabrication of crown and bridge restorations. *Australian Dental Journal* 56, no. 1 (2011):97–106. doi:10.1111/j.1834-7819.2010.01300.x
129. Sfondrini, M. F., Gandini, P., Malfatto, M., Di Corato, F., Trovati, F., Scribante, A., Computerized casts for orthodontic purpose using powder-free intraoral scanners: accuracy, execution time, and patient feedback, *BioMed Research International* 23, (2018):4103232. 10.1155/2018/4103232
130. Xue, C., Xu, H., Guo, Y., Xu, Li., Dhami, Y., Wang, H., et al., Accurate bracket placement using a computer-aided design and computer-aided manufacturing-guided bonding device: an in vivo study, *American Journal of Orthodontics and Dentofacial Orthopedics* 157, no. 2 (2020):269–277. 10.1016/j.ajodo.2019.03.022
131. Brown, M. W., Koroluk, L., Ko, C. C., Zhang, K., M. Chen, Nguyen, T., Effectiveness and efficiency of a CAD/CAM orthodontic bracket system, *American Journal of Orthodontics and Dentofacial Orthopedics* 148, no. 6 (2015):1067–1074. 10.1016/j.ajodo.2015.07.029
132. Mattheos, N., Wismeijer, D., Shapira, L., Implant dentistry in postgraduate university education. Present conditions, potential, limitations, and future trends, *European Journal of Dental Education* 18, no. 1(2014):24–32. 10.1111/eje.12074
133. Marinello, C., The digital revolution in prosthodontics: can it benefit older people? *Gerodontology* 33 (2016):145–146. doi:10.1111/ger.12228
134. Zitzmann, N. U., Kovaltschuk, I., Lenherr, I. P., Dedem, P., Joda, T., Dental students' perceptions of digital and conventional impression techniques: a randomized controlled trial, *Journal of Dental Education* 81, no. 10 (2017):1227–1232. doi:10.21815/JDE.017.081
135. Callan, R. S., Palladino, C. L., Furness, A. R., Bundy, E. L., Ange, B. L. Effectiveness and feasibility of utilizing E4D technology as a teaching tool in a preclinical dental education environment, *Journal of Dental Education* 78, no. 10 (2014):1416–1423.
136. Gottlieb, R., Lanning, S. K., Gunsolley, J. C., Buchanan, J. A., Faculty impressions of dental students' performance with and without virtual reality simulation, *Journal of Dental Education* 75, no. 11 (2011):1443–1451.
137. Buchanan, J. A., Experience with virtual reality-based technology in teaching restorative dental procedures, *Journal of Dental Education* 68, no. 12 (2004):1258–1265.
138. Raigrodski A. J., Chiche G. L., The safety and efficiency of anterior ceramic fixed partial dentures: a review of the literature, *Journal of Prosthetic Dentistry* 86, no. 5 (2001):520–525. 10.1067/mpr.2001.120111

CHAPTER 11

Spectroscopic Analysis of Dental Implants

Nebiha Gozde Ispir

11.1 INTRODUCTION

Knowledge of substances is a very important issue for treatment of patients in dentistry. Many materials such as fillings, prostheses, root canal irrigation materials, and root canal fillings are used for various dental treatments of patients. Knowing the properties of the materials used is at the forefront of successful patient treatment. In addition, these materials must have ideal properties for the success of treatments. In this regard, materials are developed by research communities consisting of many experts such as dentists, medical doctors, and engineers and examined and developed with scientific methods, in line with many hypotheses and problems.

In order to solve the problems in the materials used today, it is necessary to understand the origin and complexity of the interactions between materials (Lach et al. 2020).

It is advocated that in the research and development of materials, materials should be evaluated qualitatively and quantitatively, easily and non-destructively, and methods that can tolerate the wide range of substrate types to be used for evaluation should be used (Lach et al. 2020). There is no method that meets all these features, but also it has been stated that spectroscopic methods can meet most of these features (Lach et al. 2020). Spectroscopic techniques are used together with microscopic techniques

DOI: 10.1201/9781003506805-11

in developing dental materials and provide information about the chemical structure of dental materials (Kaczmarek et al. 2021).

The implant is one of the most used materials in the last ten years for treating edentulous areas in dentistry. Spectroscopic methods are used in recent studies for the development of implants. In this section, information about dental implants will be given, and spectroscopic methods used in recent studies on implants will be mentioned.

11.2 DENTAL IMPLANT

The aim of modern dentistry is to preserve speech, chewing, and aesthetics, which are the functions of the stomatognathic system, and to restore loss of function due to any reason. For this purpose, dental implant application is the most commonly used treatment method in recent times to eliminate tooth loss, which is one of the most important elements of the stomatognathic system.

Dental implants are produced from various biomaterials placed into the jawbone to be used as artificial tooth roots to make a fixed or removable prosthesis.

The historical basis of dental implants dates back to the Mayans. However, in 1809, a gold, tube-shaped dental implant was placed into a freshly extracted tooth socket, and a tooth-like crown was placed on top. However, this dental implant has been reported to fail and develop gingivitis (Rao and Bhat 2015).

An iridium-platinum implant soldered with gold was produced by Greenfield in 1913 (Rao and Bhat 2015). The Stock brothers applied the screw parts used in the hip bone, made of vitallium, an alloy of chrome and cobalt, as endosteal for human teeth and canine teeth (Abraham 2014). Implants with a grooved cylindrical structure, smooth margins neighboring with gingiva, and a healing cap were produced, introduced, and patented by Dr. Adams as endosseous implants in 1938 (Abraham 2014). To date, dental implants have been developed and restructured through various studies to meet the needs of individuals, taking into account their features of restoring missing tooth functions and being biocompatible. The development of implants continues to enable the stomatognathic system to fully function.

11.2.1 Materials Used in Dental Implants

It is stated in the literature that the first implant material used in implant production was marine animal shells, which were largely composed of hydroxyapatite. The materials used in the research and development process of the implant to date are as follows (Oza, Parikh, Duseja, and Agrawal 2020):

- Metals and their alloys
    a. Gold
    b. Cobalt-Crom alloys
    c. Stainless-Steel
    d. Niobium
    e. Tantalum
    f. Titanium and titanium 6-aluminum 4-vanadium
- Polymer-based materials
    a. Polymethylmethacrylate
    b. Polytetrafluoroethylene
    c. Polyethylene
    d. Polyamide
    e. Polyurethane
- Ceramics
    a. Zirconium oxide
    b. Hydroxyapatite
    c. Tricalcium Phosphate
    d. Calcium Phosphate
    e. Aluminum oxide
    f. Carbon silicon
    g. Bioglass

11.2.2 Properties Required in Materials Used for Dental Implants

The materials used in dental implants must have the following properties (Duymuş and Güngör 2013):

- It should be biocompatible with hard and soft tissues.
- It should be resistant to corrosion.

- It must have sufficient resistance against mechanical, functional and thermal stresses.
- Its surface must be coated and shaped.
- It should be easy to make and apply.

11.2.2.1 Titanium and Titanium Alloys

Today, the most used implant materials are titanium (Ti) and titanium alloys due to their excellent biocompatibility and superior mechanical properties. The reason for this can be explained by the fact that titanium is passive in the liquid environment. This feature ensures that the surface of titanium and the passive film layer formed on it remain stable in biological systems with mechanical and chemical properties (TEKİN et al. 2016). In addition, the implant material must be resistant to corrosion. The corrosion resistance of a metal is measured by its ability to oxidize quickly, the strong adhesion of the formed oxide to the substrate, and the density and protectiveness of the oxide layer formed. When the protection of titanium's oxide layer was evaluated with the Pilling–Bedworth ratio (PBR), this ratio was found to be 1.76 for titanium (Hatakeyama, Ida, Sekido, and Yoshimi 2020). The fact that PBR is less than 1 indicates that the oxide is not porous and does not have protective properties because it takes up little space in the metal. If this ratio is greater than 2, the oxide will occupy a large space and break away from the surface, leaving the surface exposed and not protecting the metal. The protection of titanium oxide (TiO_2) layer has an ideal PBR ratio and is resistant to corrosion (Hatakeyama, Ida, Sekido, and Yoshimi 2020).

TiO_2 exists in nature in three crystal polymorphic phases: anatase (tetragonal), brookite (orthorhombic), and rutile (tetragonal). The rarest phase is the Brookite phase. This phase is larger than the other phases and has eight TiO_2 groups per crystalline unit cell. This is followed by the anatase phase with four groups of TiO_2 per unit cell and the rutile phase with two groups of TiO_2 per unit cell (Xia, Lindahl, Lausmaa, and Engqvist 2011). The phases produced synthetically today are rutile and anatase. Whether it is in the rutile or anatase phase is related to the acidity of the electrolyte used. The surface tension of these phases is different from each other. Having only the rutile phase on the surface provides hydrophobic properties while having a mixture of rutile and anatase provides hydrophilic properties. The hydrophilic feature facilitates the adhesion of physiological fluids and proteins, hard and soft tissue cells (TEKİN et al. 2016; F Rupp et al. 2012).

The mechanical properties of pure titanium can also be improved by trace elements such as carbon, oxygen, nitrogen, hydrogen, and iron. These elements are found in increasing concentrations as grade I, grade II, grade III, and grade IV (Niinomi 2008).

Two titanium alloys are commercially used in dentistry for the construction of dental implants. These are Ti-6Al-4V and Ti-6Al-4V-ELI (extra-low interstitial alloy), which are formed by adding four grades of pure titanium, aluminum (Al), and vanadium (V) elements. The mechanical and physical properties of pure titanium, which consists of a mixture of titanium and oxygen, may differ depending on the oxygen residues in the metal. In addition, the strength of titanium alloys can be increased by heat treatments (Oza, Parikh, Duseja, and Agrawal 2020).

11.2.2.2 Osseointegration

In a study conducted by Branemark in 1952, titanium chambers were placed inside the bone to examine blood flow in rabbit femurs (Brånemark 1983). Over time, it has been observed that this bone clings tightly to the titanium chambers, and the titanium cannot be removed (Brånemark 1983). Thus, the concept of 'osseointegration' emerged in this study, albeit by chance (Brånemark 1983). Osseointegration is defined as a direct structural and functional connection between living bone and the surface of a load-bearing implant (Abraham 2014).

After the dental implant is placed in the jawbone, a series of biological events occur at the bone-implant interface. When the implant surface is placed into the bone, it interacts with the first blood cells. Clot (fibrin) is formed by the interaction of blood cells with the implant. The resulting fibrin functions as a scaffold for the migration (osteoconduction) of bone-forming cells (osteogenic) and the differentiation of these cells (osteoinduction). Osteogenic cells form osteoid tissue and new trabecular bone. Thus, osseointegration, a structural and functional connection, is achieved by turning into lamellar bone in direct contact with most of the implant surface (Mavrogenis, Dimitriou, Parvizi, and Babis 2009).

Osseointegration is prioritized for long-term use and success of the implant. The degree of stability achieved during osseointegration of a dental implant depends on the morphological characteristics of the jawbone or implant (Carmo Filho et al. 2018; Elias, Rocha, Nascimento, and Coelho 2012; Herrero-Climent et al. 2018; Shah, Thomsen, and Palmquist 2018).

In addition, it has been stated that the material surface and coating type have the greatest impact on osseointegration success (Lach et al. 2020).

11.2.3 Surface and Coating in Dental Implant Materials

It is important that the time for osseointegration is short for individuals to be able to fulfill their stomatognathic functions and for their comfort. The time taken for osseointegration after the dental implant is placed in the jawbone is called the healing period. Dental implant surfaces are modified in order to shorten this healing time. Various methods are used to modify. These methods include surface coating, acid etching, machining and sandblasting, electrochemical methods (anodic oxidation), vacuuming, and thermal and laser treatments (Frank Rupp et al. 2018). The aim of these processes is to obtain surface roughness. It has been found that the resulting surface roughness controls the growth and metabolic effect of osteoblasts that provide bone formation. It has been stated that surface roughness directly enables the production of transforming growth factor-beta, which increases osteoblast release (Boyan et al. 2003). Thus, the healing time is tried to be shortened by surface roughening processes.

11.2.4 Current Developments and Spectroscopic Analyses in Dental Implants

The effects of implant geometry, material type, and implant coating on osseointegration have been the subject of many studies. In addition, it has been stated that the material and coating type have the greatest impact on the success of osseointegration. Evaluation of material and coating properties has been carried out using many spectroscopic methods (Lach et al. 2020). These methods are shown in Figure 11.1.

Recently, the most used spectroscopy methods in research and development of dental implants are Raman spectroscopy, Fourier transform infrared spectroscopy (FT-IR), and X-ray spectroscopy.

Although the basic biological role of TiO_2 in osseointegration is known, it has been stated that there is limited information about the structure of TiO_2, especially about the products available on the market (Gaintantzopoulou, Zinelis, Silikas, and Eliades 2014). In their study, they aimed to identify TiO_2 phases on the root surface of 11 types of titanium dental implants applied to various surface treatments available on the market, using micro-Raman spectroscopy (Gaintantzopoulou, Zinelis, Silikas, and Eliades 2014). Raman peaks were examined by classifying them into three groups: implants with a maximum peak scale of <800 au,

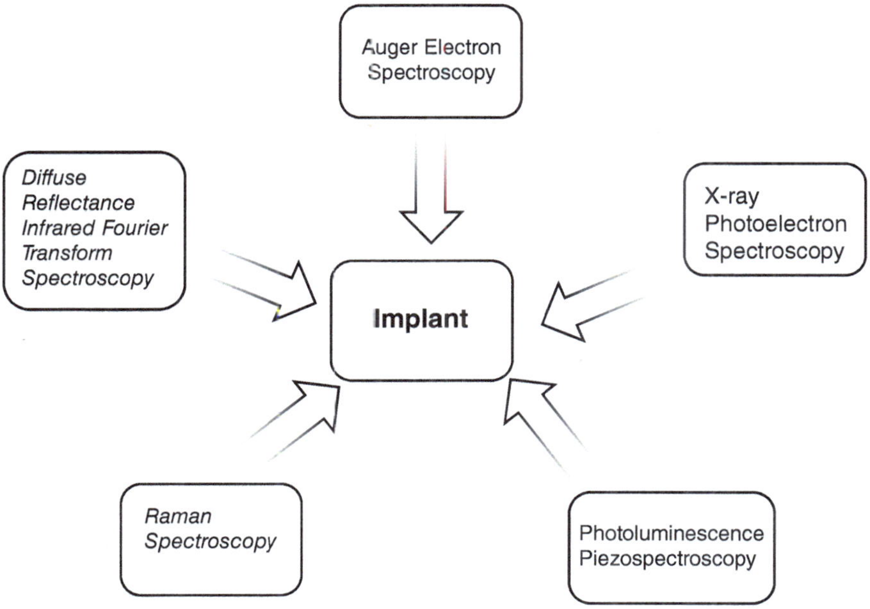

FIGURE 11.1 Spectroscopic methods used in implant studies.

implants with a maximum peak scale of 1250 to 6000 au, and implants with a maximum peak scale of >35000 au. As a result of this study, anatase was detected in only two implants, and it was shown that TiO_2 in commercially available dental implants can be found in anatase, rutile, and amorphous phases. Considering the importance of these phases in terms of osseointegration, it has been advocated that the TiO_2 phases found in the surface layers of implants on the market should be mandatory identification and evaluation by manufacturers (Gaintantzopoulou, Zinelis, Silikas, and Eliades 2014).

One of the current methods is anodization, which changes the surface of dental implants and creates a nanotopography that supports cellular activity and has the ability to interact with fluid and bone tissue. Atomic Force Microscopy (AFM), Scanning Electron Microscopy (SEM), X-ray dispersive energy spectroscopy (EDS), and Raman spectroscopy were used for implant surface characterization in a study aimed at optimizing osseointegration in dental implants through anodization, anatase phase, and achieving controlled nanotopography (Villaça-Carvalho et al. 2021). In the study, some of the titanium implants were sandblasted, some were anodized, and some were used as a control group without any extra surface treatment. It has been stated that according to the results of EDS and

Raman spectroscopy analysis, the chemical and structural composition of the titanium oxide film in the anodized implant group showed changes favorable to osteogenic activity (Villaça-Carvalho et al. 2021).

As an alternative to titanium implants, the use of polyetheretherketone, which has bone-like hardness, good fatigue resistance, high chemical resistance, and almost no toxicity, is recommended. However, polyetheretherketone is a bioinert material (Ren, Sikder, Lin, and Bhaduri 2018). The bioinert nature of a material is a disadvantage for osseointegration. The most used method to overcome the bioinert property is biofunctionalization, which is surface modification with bioactive molecules, peptides, or growth factors (Wan et al. 2019; Yakufu et al. 2020). In a study, biofunctionalization of polyetheretherketone was performed using osteogenic growth peptide (OGP) (Yakufu et al. 2020). X-ray photoelectron spectroscopy (XPS) and FT-IR spectroscopy were used to characterize the surface properties of biofunctionalized polyetheretherketone. To detect chemical groups, FT-IR spectra were applied in the wavelength range from 600 to 4000 cm^{-1} with a resolution of 2 cm^{-1}. The presence of OGP on the surfaces of polyetheretherketone was confirmed by XPS. As a result of this study, it was emphasized that covalently bonded OGP and polyetheretherketone may have great potential for orthopedic and dental applications (Yakufu et al. 2020).

As a result of microbial infections, a disease called peri-implantitis may occur involving the dental implant and surrounding tissues. Peri-implantitis causes dental implant failure. Antimicrobial agents are used for this condition, and the first choice is metronidazole. Considering this situation that causes implant failure, in a study conducted, Ti implant surfaces were coated with metronidazole with the help of the adhesive property of dopamine, and its antibacterial properties were evaluated (Fu et al. 2019). In the study, contact angle measurements, scanning electron microscopy, ATR-FTIR spectroscopy, and in vitro anti-biofilm formation assay, were applied for surface characterization of samples. The chemical components of the samples were recorded on the FTIR spectrometer, and ATR spectra were prepared under 4 000–675 cm^{-1} range, 4 cm^{-1} resolution, and room temperature conditions. The chemical components of the samples were recorded on the FTIR spectrometer, and ATR spectra were prepared under 4 000–675 cm^{-1} range, 4 cm^{-1} resolution, and room temperature conditions. ATR spectra of water were extracted from the obtained spectra. According to the spectra data of the analysis, it was shown that metronidazole was immobilized on dopamine-coated Ti. As a result of this study, a new metronidazole-modified Ti implant with antibacterial activity, which allows metronidazole to adhere to the surface with its dopamine adhesive feature, has been proposed (Fu et al. 2019).

In addition, a recent study suggested applying the electron paramagnetic resonance (EPR) method to evaluate osseointegration processes and describe biochemical dynamics (Kalinnikova et al. 2022). In this study, implants with different macro- and microstructure surfaces at the level of stable free radicals were used. The EPR method was used to describe the assimilation processes and biochemical dynamics occurring during dental implantation. According to EPR signals, it was stated that there were no paramagnetic centers in pure hydroxyapatite and the presence of impure metal ions (Mn, Fe, Cu, etc.) in the crystal structure of the samples. It has been emphasized that the EPR method, with its high sensitivity and spectral resolution, enables the determination of the most suitable implantation system and the innovative surface type with a specific morphology for the stimulation of osteoregenerative properties (Kalinnikova et al. 2022).

11.2.5 Electron Paramagnetic Resonance

The spectroscopic method used in the examination of biological substances as biogenic or synthetic is electron paramagnetic resonance (EPR) (Kalinnikova et al. 2022). Another name for EPR is electron spin resonance.

EPR was first reported by Yvgeny Zavoisky in 1945. Yvgeny Zavoisky applied radio frequency with a frequency of 133 MHz in the perpendicular direction to $CuCl_2 \cdot 2H_2O$, which he placed in an inhomogeneous magnetic field, with the device he designed for his doctoral thesis study. As the result of this application, an absorption curve of approximately 41 G magnetic field was observed (Zavoisky 1945). At about the same time, both the spectrometer and its theory were introduced by Brebis Bleinay at Oxford University (Baker 2007). Since the first observations, EPR spectroscopy has also been applied to paramagnetic biological materials. The EPR method has proven useful in the study of biological objects as it is sensitive to detect free radicals (Kalinnikova et al. 2022; Hawkins and Davies 2014).

With EPR spectroscopy, the signals formed by the response of the material as a result of stimulating the material to be examined with electromagnetic waves in the microwave frequency region in a suitable physical environment are evaluated (Pilbrow and Hanson 1993).

If we simply talk about the basis of the signals formed in traditional EPR spectroscopy: Each electron has a magnetic moment and spin quantum number. Examining a substance with EPR spectroscopy is achieved by changing the direction of the magnetic moment of an unpaired electron in the relevant atom or molecule. If a substance with a magnetic moment is exposed to a magnetic field, a coupling energy occurs. In other words, when the microwave frequency and the resonance frequency are equal, that is, when resonance occurs, a suitable spin system absorbs energy. After the impact, the spins of the electrons are directed upward or downward due to the absorbed energy. Spins directed up/down can be directed in the opposite direction by electromagnetic radiation. In this case, these spins, which are directed in the opposite direction, want to return to their previous state. For this reason, they release the energy they have on them. Thus, the released energy becomes visible as a spectrum with the EPR device (Can 2011).

11.3 CONCLUSION

Development studies are carried out on dental implants to ensure long-term healthy use and functions similar to natural teeth. These development studies continue without ignoring the satisfaction of individuals by meeting their needs and desires as they use implants. In addition, no definitive conclusion can be reached regarding the kinetics or dynamics of osseointegration with the jawbone structure, which indicates the success of an implant. Studies in implant development are multidisciplinary studies, and in these studies, there is a need for tools that can evaluate the properties of samples without the need for special pre-treatment or preparation and can be used both in vitro and in vivo. These needs can be met by spectroscopic methods, especially in modern implantology, which is responsible for the development and success of the implant surface and its interaction with the surrounding tissue.

REFERENCES

Abraham, C. M. 2014. "A brief historical perspective on dental implants, their surface coatings and treatments." *The Open Dentistry Journal* 8: 50–55. 10.2174/1874210601408010050

Baker, M. 2007. 'Brebis Bleaney.' *Physics Today* 60 (3): 80–81. 10.1063/1.2718768

Boyan, Barbara D., S. Lossdorfer, L. Wang, G. Zhao, C.H. Lohmann, D.L. Cochran, and Z. Schwartz. 2003. "Osteoblasts generate an osteogenic microenvironment when grown on surfaces with rough microtopographies." *European Cells & Materials* 6 (22): e7.

Brånemark, P. I. 1983. "Osseointegration and its experimental background." *The Journal of Prosthetic Dentistry* 50 (3): 399–410. 10.1016/s0022-3913(83)80101-2

Can, Canan. 2011. "Bazı tıbbi ve biyolojik bileşiklerde ışınlamayla oluşan hasarlar." Selçuk Üniversitesi, Yayımlanmış yüksek lisans tezi (Thesis) https://hdl.handle.net/20.500.12395/2167

Carmo Filho, Luiz Carlos Do, Raissa Micaella Marcello-Machado, Eduardo Dickie De Castilhos, Altair Antoninha Del Bel Cury, and Fernanda Faot. 2018. "Can implant surfaces affect implant stability during osseointegration? A randomized clinical trial." *Brazilian Oral Research* 32: e110.

Duymuş, Zeynep Yeşil, and Hasan Güngör. 2013. "Dental implant materyalleri." *Atatürk Üniversitesi Diş Hekimliği Fakültesi Dergisi* 23 (1): 145–152.

Elias, Carlos Nelson, Felipe Assis Rocha, Ana Lucia Nascimento, and Paulo Guilherme Coelho. 2012. "Influence of implant shape, surface morphology, surgical technique and bone quality on the primary stability of dental implants." *Journal of the Mechanical Behavior of Biomedical Materials* 16: 169–180.

Fu, Dongjie, Yi Lu, Shaohuai Gao, Youjian Peng, and Haiying Duan. 2019. "Chemical property and antibacterial activity of metronidazole-decorated Ti through adhesive dopamine." *Journal of Wuhan University of Technology-Materials Science Edition*. 34: 968–972.

Gaintantzopoulou, Marianna, Spiros Zinelis, Nick Silikas, and George Eliades. 2014. "Micro-Raman spectroscopic analysis of TiO2 phases on the root surfaces of commercial dental implants." *Dental Materials* 30 (8): 861–867.

Hatakeyama, Tomotaka, Shuntaro Ida, Nobuaki Sekido, and Kyosuke Yoshimi. 2020. "Significant improvement of the oxidation resistance of MoSiBTiC-based multiphase alloys by Ti enrichment." *Corrosion Science* 176: 108937.

Hawkins, Clare L., and Michael J. Davies. 2014. "Detection and characterisation of radicals in biological materials using EPR methodology." *Biochimica et Biophysica Acta (BBA) - General Subjects* 1840 (2): 708–721.

Herrero-Climent, Mariano, Manuel M. Romero Ruiz[a], Pedro Lazaro Calvo, José Vicente Ríos Santos, Roman A Perez, and Francisco Javier Gil Mur. 2018. "Effectiveness of a new dental implant bioactive surface: histological and histomorphometric comparative study in minipigs." *Clinical Oral Investigations* 22: 1423–1432.

Kaczmarek, K., A. Leniart, B. Lapinska, S. Skrzypek, and M. Lukomska-Szymanska. 2021. "Selected spectroscopic techniques for surface analysis of dental materials: A narrative review." *Materials (Basel)* 14 (10). 10.3390/ma14102624

Kalinnikova, Elena, Margarita Sadovnikova, Alexander Rodionov, Fadis Murzakhanov, and Peter Grishin. 2022. "Analysis of the osseointegration process of dental implants by electron paramagnetic resonance: An in vivo study." *Dentistry Journal* 10 (2): 28.

Lach, S., P. Jurczak, N. Karska, A. Kubiś, A. Szymańska, and S. Rodziewicz-Motowidło. 2020. "Spectroscopic Methods Used in Implant Material Studies." *Molecules* 25 (3). 10.3390/molecules25030579

Mavrogenis, A.F., Rozalia Dimitriou, Javad Parvizi, and George C. Babis. 2009. "Biology of implant osseointegration." *Journal of Musculoskeletal & Neuronal Interactions* 9 (2): 61–71.

Niinomi, Mitsuo. 2008. "Metallic biomaterials." *Journal of Artificial Organs* 11: 105–110.

Oza, Unnati, Hiral Parikh, Shilpa Duseja, and Charu Agrawal. 2020. "Dental implant biomaterials: a comprehensive review." *International Journal of Dental Research* 5: 87–92.

Pilbrow, John R., and Graeme R. Hanson. 1993. "[11] Electron paramagnetic resonance." In *Methods in enzymology*, 330–353. Elsevier.

Rao, BH Sripathi, and Vidya S. Bhat. 2015. "Dental implants: A boon to dentistry." *Archives of Medicine and Health Sciences* 3 (1): 131–137.

Ren, Y., P. Sikder, B. Lin, and S. B. Bhaduri. 2018. "Microwave assisted coating of bioactive amorphous magnesium phosphate (AMP) on polyetheretherketone (PEEK)." *Materials Science & Engineering. C, Materials for Biological Applications* 85: 107–113. 10.1016/j.msec.2017.12.025

Rupp, F., M. Haupt, M. Eichler, C. Doering, H. Klostermann, L. Scheideler, S. Lachmann, C. Oehr, H.P. Wendel, and E. Decker. 2012. "Formation and photocatalytic decomposition of a pellicle on anatase surfaces." *Journal of Dental Research* 91 (1): 104–109.

Rupp, F., L. Liang, J. Geis-Gerstorfer, L. Scheideler, and F. Hüttig. 2018. "Surface characteristics of dental implants: A review." *Dental Materials* 34 (1): 40–57.

Shah, F.A., P. Thomsen, and A. Palmquist. 2018. "A review of the impact of implant biomaterials on osteocytes." *Journal of Dental Research* 97 (9): 977–986.

Tekin, Uğur, Mehmet Ali Güngör, M. Erhan Çömlekoğlu, Akin Aladağ, and Mine Dündar Çömlekoğlu. 2016. "Dental İmplantların Yüzey Yapıları." *Turkiye Klinikleri Journal of Dental Sciences* 22 (1): 55–62.

Villaça-Carvalho, M. F. L., J. C. R. de Araújo, J. M. Beraldo, R. F. D. Prado, M. E. L. Moraes, L. R. C. Manhães Junior, E. N. Codaro, H. A. Acciari, J. P. B. Machado, N. N. Regone, A. O. Lobo, F. R. Marciano, and L. M. R. Vasconcellos. 2021. "Bioactivity of an Experimental Dental Implant with Anodized Surface." *Journal of Functional Biomaterials* 12 (2): 39. 10.3390/jfb12020039

Wan, Teng, Linlong Li, Min Guo, Zixue Jiao, Zongliang Wang, Yoshihiro Ito, Yizao Wan, Peibiao Zhang, and Qinyi Liu. 2019. "Immobilization via polydopamine of dual growth factors on polyetheretherketone: improvement of cell adhesion, proliferation, and osteo-differentiation." *Journal of Materials Science* 54 (16): 11179–11196.

Xia, Wei, Carl Lindahl, Jukka Lausmaa, and Håkan Engqvist. 2011. "Biomimetic hydroxyapatite deposition on titanium oxide surfaces for biomedical application." *Advances in Biomimetics* 20: 429–452.

Yakufu, Maihemuti, Zongliang Wang, Yu Wang, Zixue Jiao, Min Guo, Jianguo Liu, and Peibiao Zhang. 2020. "Covalently functionalized poly (etherether-ketone) implants with osteogenic growth peptide (OGP) to improve osteogenesis activity." *RSC Advances* 10 (17): 9777–9785.

Zavoisky, E. 1945. "Spin-magnetic resonance in paramagnetics." *Journal Physics USSR* 9: 211–245.

CHAPTER 12

Spectroscopic Methods to Determine the Toxicity of Biomaterials Used in Dentistry

Ilkay Peker

12.1 INTRODUCTION

Several biomaterials are used for a wide variety of purposes in dentistry. The effort to find the best method and material to rehabilitate missing teeth and restore dental structures is a process that continues today. New materials are frequently added to the existing materials on the market. Dental materials remain in direct contact with vital tissues such as tooth structures, bone, and oral mucosa for a long time. Therefore, the harmful effects of these materials on oral tissues should be evaluated before clinical use.

Biomaterial or biocompatible material is defined as a material that causes a minimum level of response in the tissue and helps restore the function of the tissue or organ after the non-toxic material is placed into the body to treat a tissue or organ that has lost its function (DF 1992). Toxicity refers to the degree of damage that a substance can cause in the organism (Ranjeet Ajit Bapat et al. 2021). It is crucial for the long-term success of dental treatments to evaluate all materials used in the human body regarding biocompatibility and their physical and mechanical properties and to pay attention to selecting reliable materials. However, knowledge regarding the adverse effects of biomaterials used in dentistry is

DOI: 10.1201/9781003506805-12

crucial for the safety of patients and dental personnel who come into contact with these materials.

Advancements in spectroscopic techniques allow detailed chemical structure and surface analysis of the materials (Al-Ahdal, Silikas, and Watts 2015). Dental staff face the risk of being exposed to the adverse effects of these materials as much as dental patients (Ranjeet Ajit Bapat et al. 2021). Dental personnel are at higher risk than patients for some dental biomaterials (resin, rubber-based materials, etc.). They may cause adverse reactions on the hands and fingertips of dental personnel due to long-term exposure. This chapter aims to review spectroscopic methods to determine the toxicity of biomaterials used in dentistry.

12.2 PROPERTIES OF DENTAL BIOMATERIALS

Many factors, such as the location of dental material, placement in soft or mineralized tissue, direct or limited contact with the oral epithelium such as dental implants, direct contact with bone, tissue fluids, blood, and saliva, or through any barrier such as the dentin and enamel barrier, can affect the material. It plays an active role in developing biological responses (Anusavice 2003). The duration of the material remaining in the body also closely affects the biological response. The biological effects of materials that remain in the mouth for a short time differ from those that remain in the mouth for longer (Anusavice 2003).

12.2.1 Factors Affecting the Properties of Dental Materials

Although biomaterials in body parts remain at the appropriate temperature and have a constant chemical structure, the oral environment is exposed to extreme temperature and pH changes and foods with different chemical structures. Extreme temperature changes from ice cream (0°C) to hot coffee (90°C) can lead to compatibility problems such as thermal expansion, changes in mechanical properties, or bonding failure (St John 2007).

Many essential properties are considered, such as its chemical composition, surface structure, behavior against liquids, flexibility, wear resistance, and electrical charge. Biomaterials used in dentistry should be resistant to physicochemical factors in the mouth, not be affected by biophysical changes, be able to imitate natural tooth structures, have stable mechanical properties, be easy to process, have suitable aesthetic properties, and be a reasonable price (Wojda 2015; Zhang and Kelly 2017; Della Bona et al. 2021). At the same time, properties of the dental material, such as surface structure, roughness, chemical composition, and reactivity, should be evaluated (Wojda 2015; Guglielmotti, Olmedo, and Cabrini 2019; Zhang and Kelly 2017; Sterzenbach et al. 2020; Revilla-León et al. 2021; Ercoli and Caton 2018). First, a biofilm forms on the biomaterial placed in the oral cavity (Wojda 2015; Eick 2021; Sterzenbach et al. 2020). To prevent microleakage and biofilm formation, it is necessary to know and examine the structural and chemical components of the material (Wojda 2015; Koutouzis 2019; Bhadila et al. 2021; Tettamanti et al. 2017).

12.2.2 Classification of Dental Biomaterials

Dental biomaterials can be classified into five main headings commonly used in dentistry. This classification and the frequently used materials for each group are presented in Figure 12.1.

Application areas of dental biomaterials in the mouth are presented in Figure 12.2.

12.3 TOXICITY OF DENTAL BIOMATERIALS

Side effects and adverse reactions of dental biomaterials are rare and often not severe. However, prolonged contact and exposure to living tissues should be considered. Biocompatibility of dental materials is a complex issue that includes biological factors, patient risk factors, clinical experience, and engineering issues (Ranjeet Ajit Bapat et al. 2021).

Dental biomaterials can cause undesirable biological and immunological reactions, the most common adverse effects being allergic reactions (Ranjeet Ajit Bapat et al. 2021). Allergic reactions usually occur with swelling, rash, urticaria, and runny nose but can also be life-threatening, such as anaphylaxis, laryngeal edema, and cardiac arrhythmias (Ranjeet Ajit Bapat et al. 2021; Karabucak and Stoopler 2007). Clinical symptoms are mostly rash, dry mouth, burning in the mouth, pain, cheilitis, erosion of the oral mucosa, and stomatitis (Ditrichova et al. 2007). In addition, different dental biomaterials have been observed to cause cell-mediated and humoral immune reactions (Karabucak and Stoopler 2007). Contact allergic reactions occur due to the release of allergen components, cytotoxic metallic ions, or formaldehyde (Mjör 1992; Ranjeet A Bapat et al. 2019). Dental staff face the risk of being exposed to the adverse effects of these materials as much as dental patients (Ranjeet Ajit Bapat et al. 2021).

It has been reported that amalgam has a higher potential to cause oral lichenoid reactions than other dental biomaterials. Therefore, it should be used with caution in patients with a history of allergies. It is also stated

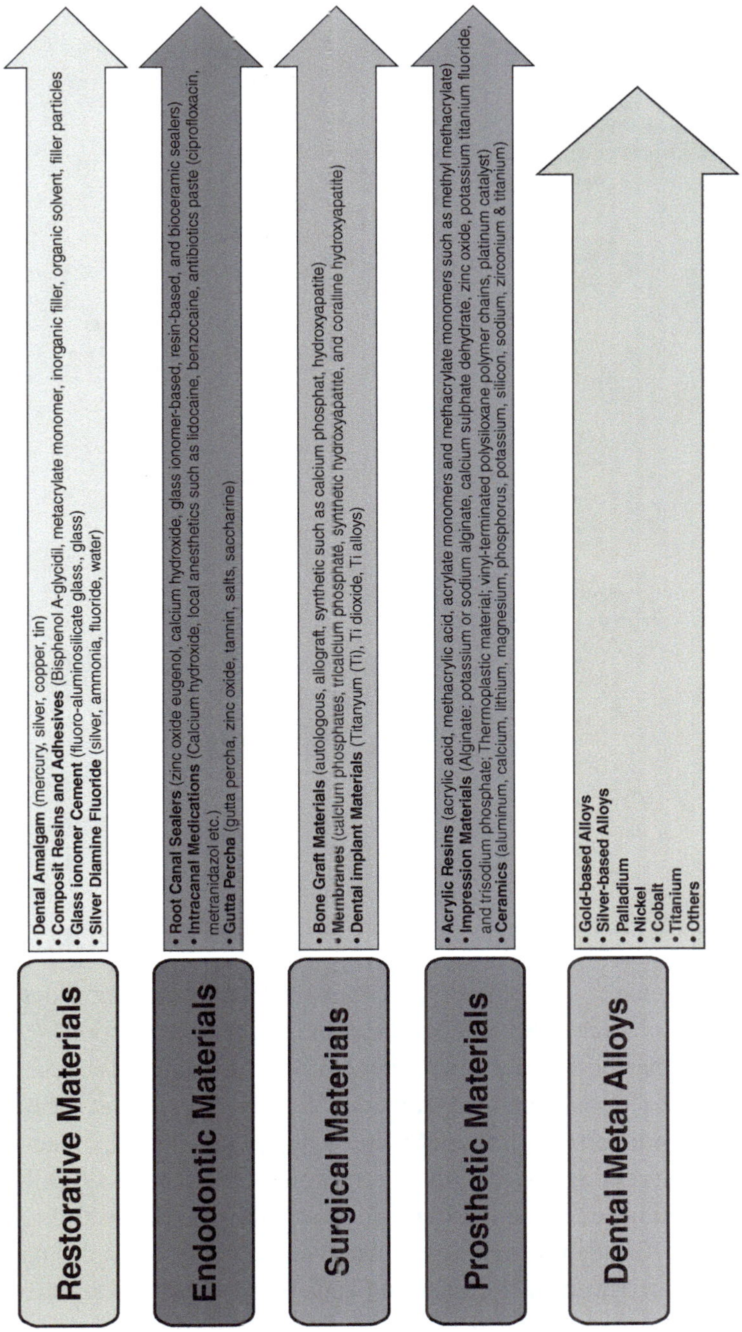

FIGURE 12.1 Biomaterials commonly used in dentistry.

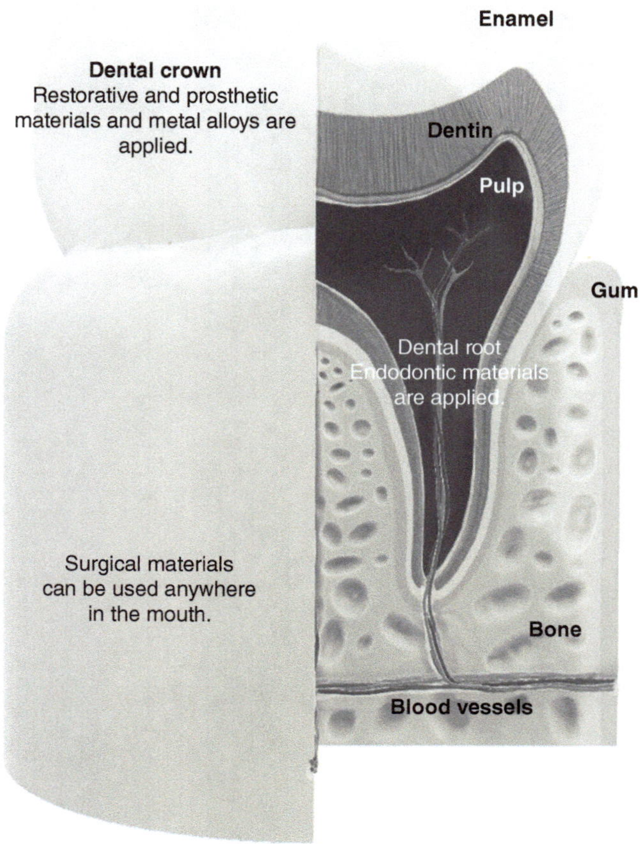

FIGURE 12.2 Application areas of dental biomaterials in the mouth.

that glass ionomer cement, composite resins, and dental casting alloys that release metal ions may cause allergic and oral lichenoid reactions (Shahi et al. 2019). It has been determined that resin-based materials cause skin and mucosal reactions in patients and dentists. It is thought that the unpolymerized monomers in these materials cause allergic reactions. Accordingly, dermatoses are the most common (Shahi et al. 2019). Allergic reactions to titanium implants occur due to metal ion release and corrosion products. These may include edema, rash, eczema, atopic dermatitis, necrosis, and urticaria in the skin and mucosa. Individuals sensitive to other metals have a higher risk of being allergic to titanium (Shahi et al. 2019).

12.3.1 Biocompatibility Evaluations of Dental Biomaterials

Dental biomaterials must pass a series of tests before being used in clinical practice. Guidelines have been published by the American Dental Association (ADA), the Food and Drug Administration (FDA), and the International Organization for Standardization (ISO) regarding the suitability of existing and new materials before their use in dentistry (Anusavice 2003; Schmalz, Arenholt-Bindslev, and Schmalz 2009). ISO 7405278 is authorized for clinical testing of the biocompatibility of medical devices used in dentistry (Standard 2017). Test evaluations for dental materials are a combination and coordination of international and national standards and guidelines.

12.3.2 Toxicity Tests of Dental Biomaterials

In vitro cytotoxicity tests are initial tests used in the biocompatibility evaluation of biomaterials. After successfully completing these tests, the material is tested on animals in an environment mimicking clinical conditions. After success is achieved in in vitro cytotoxicity and animal studies, the biocompatibility of biomaterials is tested on humans. These tests are carried out under ethical conditions and by legal regulations. Biomaterials approved by the FDA can be used for clinical trials (DF 1992; Swetha et al. 2015; Schmalz, Arenholt-Bindslev, and Schmalz 2009). Restorative materials are clinically tested according to the guidance of the United States Public Health Service.

Biomaterials are tested in four stages:(Ranjeet Ajit Bapat et al. 2021)

-Phase I: The material is tested in a small group of patients to evaluate safe doses.

-Phase II: The material is tested in a larger group of patients to test for any adverse effects.

-Phase III: The material is compared using a placebo and a common treatment procedure in a large group of patients from different countries.

-Phase IV: Evaluations of the material's drug interactions, side effects, and effectiveness in a broader human population are evaluated (Ranjeet Ajit Bapat et al. 2021).

Toxicity test methods of dental biomaterials are presented in Figure 12.3.

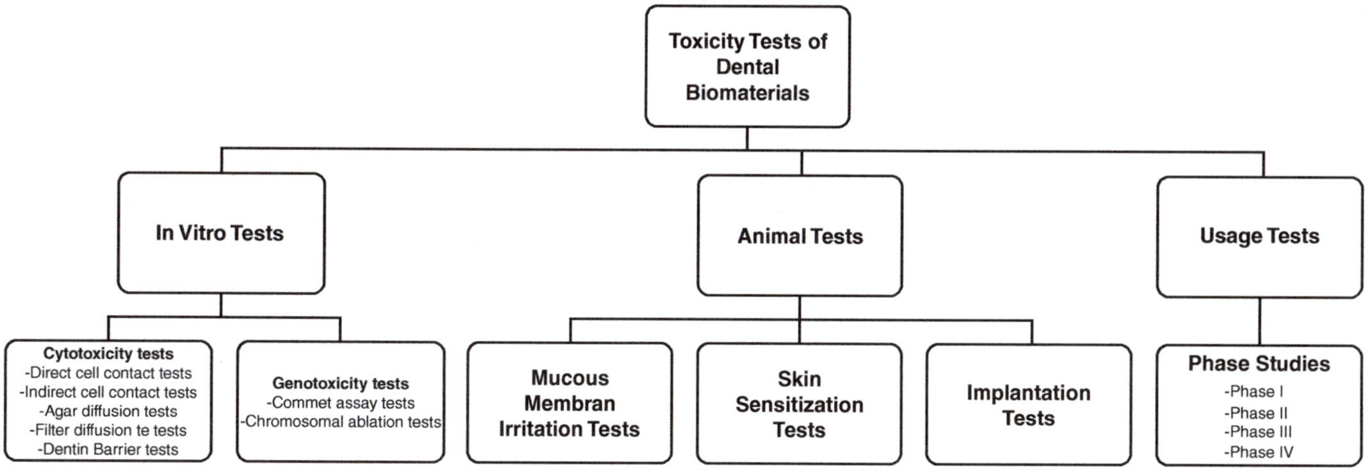

FIGURE 12.3 Classification of toxicity tests of dental biomaterials.

12.4 SPECTROSCOPIC METHODS FOR TOXICITY OF DENTAL BIOMATERIALS

Cytotoxicity tests of dental biomaterials are mostly performed by microscopic methods involving cell culture. In recent years, studies have been carried out using spectroscopic methods and microscopic methods to evaluate the cytotoxicity of dental materials. Spectroscopic methods, whose basic principle is mattering interaction with electromagnetic radiation, are used in many scientific fields by allowing detailed matter analysis. These methods allow both the evaluation of the chemical structures of materials and surface analysis. It is reported that methods such as infrared spectroscopy (IR), Fourier Transform infrared spectroscopy (FT-IR), Raman spectroscopy, ultraviolet and visible spectroscopy (UV-Vis), X-ray spectroscopy, and Mass Spectrometry (MS) are used (spectroscopic, selected spectroscopic). Additionally, studies using UV spectrometry, atomic flame spectrophotometry (AFS), and atomic absorption spectroscopy (AAS) have also been conducted to evaluate the potential cytotoxicity and biocompatibility of dental biomaterials (Halbach et al. 2008; Braga, Catirse, and Spadaro 2005; Weingart et al. 1994).

12.4.1 Atomic Absorption Spectroscopy

AAS is a method frequently used to determine elements and is based on measuring the absorption of light by atoms at the fundamental energy level. It is especially preferred for quantitative analysis of trace amounts of metals. Nearly 60 elements in dental biomaterials, such as mercury (Hg), magnesium (Mg), plumbic (Pb), copper (Cu), nickel (Ni), aluminum (Al), chrome (Cr), and cobalt (Co), with a wavelength of 193.7–852.1 nm, can be determined with this method (Welz and Sperling 2008).

Corrosion products released in dental restorations in the mouth and causing toxic reactions are 1 mg/cm per week. In an in vitro study by Jayaprakash et al., ion release from dental casting alloys was evaluated quantitatively with AAS under various pH conditions. It has been shown that very dilute concentrations of metal ions can be measured with AAS (Jayaprakash et al. 2017). AAS was used in the study to measure the amount of mercury exposed to human subjects through inhalation due to amalgam restorations. Mercury exposure has been reported to be tolerable (Berglund 1990). Another study by Halbach et al. evaluated mercury levels in human subjects' blood and urine samples with AAS after amalgam fillings removal and found similar results (Halbach et al. 2008). The presence of titanium (Ti) in the regional lymph nodes of patients with endosseous Ti implants was evaluated using the AAS method. It has been determined that Ti does not cause a foreign body reaction in the relevant lymph nodes (Weingart et al. 1994).

Flame AAS, in which the atomization process is carried out in the flame, was used to evaluate the toxicity of the zinc (Zn) and cadmium (Cd) amounts in the alginate used as an impression material (Alginate). It has been determined that alginates do not contain toxic amounts of metal (Braga, Catirse, and Spadaro 2005).

12.4.2 Ultraviolet (UV) Spectrometry

The ultraviolet and visible light (UV-Vis) absorption spectroscopy method is based on measuring the attenuation of a beam of light after passing through a sample or being reflected from a sample surface. With UV spectrometry, Engler et al. analyzed the presence and amount of residual monomers in various dental polymers, which can cause potential toxic effects. At the end of the study, they reported that the amount of residual monomer was below standard threshold values (Engler et al. 2020).

Fourier transform infrared spectroscopy, a type of infrared spectrometer, is used to determine the toxicity of dental materials. FTIR and XRD spectroscopy were used in a study evaluating the cytotoxicity of fluoride-added calcium phosphates used for remineralization purposes. In the study, these materials were found to be reliable in terms of biocompatibility (Sauro et al. 2023). X-ray spectroscopy and FTIR were used to analyze the toxicity of acrylate-based restorative composite resins filled with hydroxyapatite and silica/hydroxyapatite combination. As a result of the study, it was concluded that restorative composite resins are non-toxic and reliable biomaterials (Chadda et al. 2016).

12.4.3 Mass Spectrometry

Mass spectrometry is an analytical method based on evaporating the sample, ionizing it, and recording the resulting ions by separating them according to their mass/charge values. Mass spectrometry evaluated the toxic effects of ions released from base metal alloys exposed to *Streptococcus. mutans*, a bacterium in the oral environment. The result of the study showed that these materials have harmful effects (McGinley et al. 2013).

12.4.4 X-Ray Spectroscopy

Qualitative and quantitative analyses based on electromagnetic emission, absorption, scattering, fluorescence, and diffraction measurements can be performed with X-ray spectroscopy. There are other methods of X-ray spectroscopy, such as X-ray absorption spectroscopy (XAS), X-ray emission spectroscopy (XES), X-ray fluorescence spectroscopy (XRF), X-ray photoelectron spectroscopy (XPS), and X-ray diffraction (XRD), are used for analyses of dental materials (Wojda 2015). A study was conducted to investigate the long-term effects of these materials on teeth restored with amalgam, compomer, and glass ionomer cement. The study results showed that not only mercury was released from the amalgam, but also trace amounts of copper were released (Mass, Hassan, and Zilberman 2017). The XRF method was used in a study investigating the diffusion of amalgam components Ag, Cu, Sn, and Hg into human teeth. In the study, it was found that mercury was released from amalgam (Guerra et al. 2016).

A study was conducted using Raman spectroscopy and XRD spectroscopy to evaluate the in vitro cytotoxicity of bone graft materials obtained from synthetic, bovine, and human sources. The study's results unexpectedly showed that these materials may have a cytotoxic effect, and it was recommended that they be used cautiously (Yang et al. 2022).

12.5 CONCLUSION

Biocompatibility evaluation of various dental biomaterials is vital for patients and dental personnel. Any material must undergo numerous tests before being used safely on humans. Spectroscopic methods are successfully used in toxicity assessment by allowing surface analysis and chemical structures of dental biomaterials to be examined. The use of these methods in biocompatibility evaluation studies arouses interest from researchers.

REFERENCES

Al-Ahdal, Khold, Nick Silikas, and David C Watts. 2015. "Development of viscoelastic stability of resin-composites incorporating novel matrices." *Dental Materials: Official Publication of the Academy of Dental Materials* 31 (12): 1561–1566.

Anusavice, K.J. 2003. *Phillips' Science of Dental Materials - eBook: Phillips' Science of Dental Materials - eBook*. Elsevier Health Sciences.

Bapat, Ranjeet A, Suyog Dharmadhikari, Tanay V Chaubal, Mohd Cairul Iqbal Mohd Amin, Prachi Bapat, Bapi Gorain, Hira Choudhury, Christopher Vincent, and Prashant Kesharwani. 2019. "The potential of dendrimer in delivery of therapeutics for dentistry." *Heliyon* 5 (10): e02544.

Bapat, Ranjeet Ajit, Abhishek Parolia, Tanay Chaubal, Suyog Dharamadhikari, Anshad Mohamed Abdulla, Nasil Sakkir, Suraj Arora, Prachi Bapat, Amal M Sindi, and Prashant Kesharwani. 2021. "Recent update on potential cytotoxicity, biocompatibility and preventive measures of biomaterials used in dentistry." *Biomaterials Science* 9 (9): 3244–3283.

Berglund, A. 1990. "Estimation by a 24-hour study of the daily dose of intra-oral mercury vapor inhaled after release from dental amalgam." *Journal of Dental Research* 69 (10): 1646–1651.

Bhadila, Ghalia, Deepak Menon, Xiaohong Wang, Taissa Vila, Mary Ann S Melo, Silvia Montaner, Dwayne D. Arola, Michael D. Weir, Jirun Sun, and H.K. Hockin. 2021. "Long-term antibacterial activity and cytocompatibility of novel low-shrinkage-stress, remineralizing composites." *Journal of Biomaterials Science. Polymer edition* 32 (7): 886–905.

Braga, A.S., A.B.C.E.B. Catirse, and A.C.C. Spadaro. 2005. "Quantitative analysis of potentially toxic metals in alginates for dental use." *Revista de Ciências Farmacêuticas Básica e Aplicada* 26 (2): 125–130.

Chadda, Harshita, Sangeetha Vasudevaraj Naveen, Saktiswaren Mohan, Bhabani K Satapathy, Alok R Ray, and Tunku Kamarul. 2016. "Cytotoxic evaluation of hydroxyapatite-filled and silica/hydroxyapatite-filled acrylate-based restorative composite resins: An in vitro study." *The Journal of Prosthetic Dentistry* 116 (1): 129–135.

Della Bona, Alvaro, Viviane Cantelli, Vitor T Britto, Kaue F Collares, and Jeffrey W Stansbury. 2021. "3D printing restorative materials using a stereolithographic technique: a systematic review." *Dental Materials: Official Publication of the Academy of Dental Materials* 37 (2): 336–350.

Df, Williams. 1992. "Second consensus on definitions in biomaterials." *Biomaterial-tissue Interfaces* 10: 525–533.

Ditrichova, Dagmar, Simona Kapralova, Martin Tichy, Vlastislava Ticha, Jitka Dobesova, Eva Justova, Miroslav Eber, and Petr Pirek. 2007. "Oral lichenoid lesions and allergy to dental materials." *Biomedical Papers of the Medical Faculty of the University Palacky, Olomouc, Czechoslovakia* 151 (2): 333–339.

Eick, Sigrun. 2021. "Biofilms." *Monographs in Oral Science* 29: 1–11.

Engler, Madalena Lucia Pinheiro Dias, Jan-Frederik Güth, Christine Keul, Kurt Erdelt, Daniel Edelhoff, and Anja Liebermann. 2020. "Residual monomer elution from different conventional and CAD/CAM dental polymers during artificial aging." *Clinical Oral Investigations* 24 (1): 277–284.

Ercoli, Carlo, and Jack G. Caton. 2018. "Dental prostheses and tooth-related factors." *Journal of Clinical Periodontology* 45: S207–S218.

Guerra, M., C. Ferreira, M.L. Carvalho, J.P. Santos, and S. Pessanha. 2016. "Distribution of toxic elements in teeth treated with amalgam using μ-energy dispersive X-ray fluorescence." *Spectrochimica Acta Part B: Atomic Spectroscopy* 122: 114–117.

Guglielmotti, María B., Daniel G. Olmedo, and Rómulo L. Cabrini. 2019. "Research on implants and osseointegration." *Periodontology 2000* 79 (1): 178–189.

Halbach, S, S. Vogt, W. Köhler, N. Felgenhauer, G. Welzl, L. Kremers, T. Zilker, and D. Melchart. 2008. Blood and urine mercury levels in adult amalgam patients of a randomized controlled trial: interaction of Hg species in erythrocytes. *Environmental Research* 107 (1): 69–78.

Jayaprakash, K., K. Harish Kumar Shetty, A. Nityananda Shetty, and B.T. Nandish. 2017. "Effect of recasting on element release from base metal

dental casting alloys in artificial saliva and saline solution." *Journal of Conservative Dentistry: JCD* 20 (3): 199–203.
Karabucak, B., and E.T. Stoopler. 2007. "Root canal treatment on a patient with zinc oxide allergy: a case report." *International Endodontic Journal* 40 (10): 800–807.
Koutouzis, Theofilos. 2019. "Implant-abutment connection as contributing factor to peri-implant diseases." *Periodontology 2000* 81 (1): 152–166.
Mass, Elyiahu, Alla Hassan, and Uri Zilberman. 2017. "Long-term in-vivo effect of various restorative materials on enamel and dentin of primary molars." *Quintessence International (Berlin, Germany: 1985)* 48 (8): 633–638.
McGinley, E.L., A.H. Dowling, G.F. Moran, and G.J.P. Fleming. 2013. "Influence of S. mutans on base-metal dental casting alloy toxicity." *Journal of Dental Research* 92 (1): 92–97.
Mjör, I.A. 1992. "Problems and benefits associated with restorative materials: side-effects and long-term cost." *Advances in Dental Research* 6: 7–16.
Revilla-León, Marta, Jorge Alberto Morillo, Wael Att, and Mutlu Özcan. 2021. "Chemical composition, knoop hardness, surface roughness, and adhesion aspects of additively manufactured dental interim materials." *Journal of Prosthodontics: Official Journal of the American College of Prosthodontists* 30 (8): 698–705.
Sauro, Salvatore, Gianrico Spagnuolo, Carmela Del Giudice, Davino M Andrade Neto, Pierre BA Fechine, Xiaohui Chen, Sandro Rengo, Xiaojing Chen, and Victor P Feitosa. 2023. "Chemical, structural and cytotoxicity characterisation of experimental fluoride-doped calcium phosphates as promising remineralising materials for dental applications." *Dental Materials: Official Publication of the Academy of Dental Materials* 39 (4): 391–401.
Schmalz, Gottfried, Dorthe Arenholt-Bindslev, and Gottfried Schmalz. 2009. "Determination of biocompatibility." In: *Biocompatibility of Dental Materials*. Springer, Berlin, Heidelberg. p. 13–43.
Shahi, Shahriar, Mutlu Özcan, Solmaz Maleki Dizaj, Simin Sharifi, Nadin Al-Haj Husain, Aziz Eftekhari, and Elham Ahmadian. 2019. "A review on potential toxicity of dental material and screening their biocompatibility." *Toxicology Mechanisms and Methods* 29 (5): 368–377.
St John, Kenneth R. 2007. "Biocompatibility of dental materials." *Dental Clinics Of North America* 51 (3): 747–viii.
Standard, I. 2017. "ISO 10993-4: biological evaluation of medical devices part 4—selection of tests for interactions with blood." *International Organization for Standardization*: Geneva, Switzerland.
Sterzenbach, Torsten, Ralf Helbig, Christian Hannig, and Matthias Hannig. 2020. "Bioadhesion in the oral cavity and approaches for biofilm management by surface modifications." *Clinical Oral Investigations* 24 (12): 4237.
Swetha, B., Sylvia Mathew, B.V. S. Murthy, N. Shruthi, and S. H. Bhandi. 2015. "Determination of biocompatibility: A review." *International Dental & Medical Journal of Advanced Research* 1 (1): 1–6.
Tettamanti, L., F. Cura, C. Andrisani, M. Andreasi Bassi, J. Silvestrerangil, and A. Tagliabue. 2017. "A new implant-abutment connection for bacterial

microleakage prevention: an in vitro study." *Oral & Implantology* 10 (2): 172–180.

Weingart, D., S. Steinemann, W. Schilli, J.R. Strub, U. Hellerich, J. Assenmacher, and J. Simpson. 1994. "Titanium deposition in regional lymph nodes after insertion of titanium screw implants in maxillofacial region." *International Journal of Oral and Maxillofacial Surgery* 23 (6 Pt 2): 450–452.

Welz, B., and M. Sperling. 2008. *Atomic Absorption Spectrometry*. Wiley.

Wojda, Sylwia M. 2015. "Comparative analysis of two methods of assessment wear of dental materials." *Acta Mechanica et Automatica* 9 (2): 105–109.

Yang, Fan, Kao Li, Shi Fu, Michael Cuiffo, Marcia Simon, Miriam Rafailovich, and Georgios E Romanos. 2022. "In vitro toxicity of bone graft materials to human mineralizing cells." *Materials (Basel, Switzerland)* 15 (5): 1955.

Zhang, Yu, and J. Robert Kelly. 2017. "Dental ceramics for restoration and metal veneering." *Dental Clinics of North America* 61 (4): 797–819.

Index

Page numbers in *italic* indicate figures and Page numbers in **bold** indicate tables.

E

F

N

O

U

V

W

X

Y

Z

FSC
www.fsc.org
MIX
Papier | Fördert gute Waldnutzung
FSC® C083411

Kolibri 360 GmbH
Ferdinand-Jühlke-Straße 7
99095 Erfurt, Deutschland
produktsicherheit@kolibri360.de